Handbook of Manufacturing and Production Management Formulas, Charts, and Tables

Donald W. Moffat

PRENTICE HALL
Englewood Cliffs, New Jersey 07632

Prentice-Hall International, Inc., *London*
Prentice-Hall of Australia, Pty. Ltd., *Sydney*
Prentice-Hall Canada, Inc., *Toronto*
Prentice-Hall of India Private Ltd., *New Delhi*
Prentice-Hall of Japan, Inc., *Tokyo*
Prentice-Hall of Southeast Asia Pte. Ltd., *Singapore*
Editora Prentice-Hall do Brasil Ltda., *Rio de Janeiro*
Prentice-Hall Hispanoamericana, S.A., *Mexico*

© 1987 *by*

PRENTICE-HALL, INC.

Englewood Cliffs, N.J.

10 9 8 7 6 5 4 3 2

Dedicated
with love
to
Aunt Florence and Perk

Library of Congress Cataloging-in-Publication Data

Moffat, Donald W.
  Handbook of manufacturing and production management
formulas, charts, and tables.

  Includes index.
  1. Production management—Handbooks, manuals, etc.
I. Title.
TS155.M644 1987   670'.0202   87-11430

ISBN 0-13-379256-0

PRENTICE HALL
BUSINESS & PROFESSIONAL DIVISION
A division of Simon & Schuster
Englewood Cliffs, New Jersey 07632

Printed in the United States of America

ABOUT THE AUTHOR

DONALD W. MOFFAT is the author of the best-selling book, *Plant Engineer's Handbook of Formulas, Charts and Tables*, Second Edition (Prentice-Hall, Inc.); *Concise Desk Book of Business Finance*, Second Edition (Prentice-Hall, Inc.); *Economics Dictionary*, Second Edition (Elsevier Science Publishing Company, Inc.); and *Charts & Nomographs for Electronics Technicians and Engineers* (Tab Books Inc.), (which simplifies mathematics for industrial technicians and provides many aids for rapid evaluation of frequently used functions). His next book, to be published by Prentice-Hall, Inc., is titled *Manufacturing Engineer's Complete Guide to Project Management*.

He taught industrial management and business at Miramar College in San Diego, and was Chairman of the Mathematics Department at The Bishop's School. He also taught upper division and graduate courses in business at LaVerne College.

Before going into full-time teaching, he worked in industry for twenty-six years at various levels from construction mechanic to project engineer, production coordinator, industrial teacher, and program management. He has presented nine papers to national and international professional conferences and had over 55 articles published.

Donald W. Moffat's formal education includes a Bachelor of Science in physics from Hofstra University, graduate studies at Syracuse University and UCLA, a Master of Science in Management Science from United States International University, and a teaching credential program at University of San Diego. He is a member of Sigma Pi Sigma, National Physics Honor Society.

WHAT THIS BOOK WILL DO FOR YOU

Handbook of Manufacturing and Production Management Formulas, Charts, and Tables is specifically written for all production managers and executives who make production decisions and who don't have time to waste searching for the right answers. This compact source offers the information you need for planning, forecasting, analyzing, and testing the procedures used in operating your production plant.

This *Handbook* is *not* a textbook; it is meant for users who have training or experience in fundamental mathematics. On the other hand, you don't need to have kept up with all the mathematics you learned in college—you only need enough to make substitutions in the formulas. You'll find direct, to-the-point examples with descriptions, procedures, and explanations that are both directly and indirectly related to production.

This *Handbook* consists of formulas, tables, explanations, and examples. Because even a slight change in a formula can have a significant effect on its intended application, at least one example is included with each formula to help clarify the meaning and applications of the symbols and formulas.

Most of the formulas can be solved with a simple hand calculator; many do not need a calculator at all. A few of the formulas use exponentials, and they are easier to solve with a scientific calculator. In addition, included are computer programs that solve all the formulas in their respective chapters. The programs make it convenient to solve any formula repeatedly used, thus allowing you to investigate many sets of numbers. Some of the programs also print tables of results.

Here are just some of the many benefits this book offers.

- Formulas to help you plan schedules and use of facilities for the best use of machinery and other equipment.

- How to use mathematical simulation to test the effects of decisions by simulating months or years of operation in a short time.

- How to get the most effective use of money spent for quality control, and how to find the best position between overtesting (expensive) and insufficient testing (which damages the company image).

- Queuing calculations to help find the optimum position between facilities that wait to serve and employees or machines that wait to be served.

- How to express past patterns analytically and use them to look into the future.

- Techniques of linear programming and how it can help you make the most of your facilities.

- Measurement techniques and analysis to help pinpoint inefficient uses of labor.

- How to find the least-cost point between extremes. For example, carrying too much inventory is inefficient, and too little inventory frequently causes production shutdowns.

You'll find *Handbook of Manufacturing and Production Management Formulas, Charts, and Tables* saves you time because all the information you need is here in one volume, with computer programs to help you make the necessary calculations quickly and allow you to examine many situations in a short time.

HOW TO USE THIS HANDBOOK

This *Handbook* is carefully formatted; it first gives the symbols used in the chapter, introduces a concept, presents the formula or table, and then provides several examples.

When you find a formula or table that you wish to use, check the description to see if your application is within any given limitations. Then review the example provided with it (there is at least one example for every formula) to verify that you have the right understanding of each symbol's meaning.

The following sequence, from Chapter 1, *Key Formulas for Getting Started in the Production Process*, is typical. A formula from the group on the learning function gives the average time a person (or team, organization, etc.) takes to complete a task.

$$a_n = a_l/n^k$$

At the beginning of the chapter you find explanations of the symbols:

a_l = time to complete the task the first time

a_n = overall average unit time after completing the task *n* times

k = learning factor

n = number of times task has been completed

A further discussion of one symbol (k) is given with the formula. An example then follows in the chapter:

Example. One position in an electronic board assembly line requires that the operator insert two components, connect eleven wires, and solder nine places. The operator took 96 seconds the first time. What overall average can we expect after twenty boards?

Assume a medium value of 0.1 for k, because the operator already knows how to wire and solder; only the new sequence has to be learned.

$$a_{20} = 96/20^{0.1}$$

$$= 71 \text{ seconds}$$

When a formula uses numbers that you need to estimate, it is helpful to make three calculations. One should be your best estimate (most likely) of all the variables, and one each with the most pessimistic and the most optimistic view of all variables.

For example, the Chapter 5, *Formulas to Help Minimize Inventory Costs and Improve Ordering Procedures*, on inventory policies gives the following formula for finding the cost of the optimum inventory policy (the policy that results in the lowest cost):

$$c_{to} = \sqrt{2\ c_p\ c_i\ u\ (1 - r_u/r_i)}$$

The meanings of all symbols are explained in the chapter. Let us say that c_p is 27.8 dollars, c_i is 5.75 dollars, and u is 4,500. Our best estimate of r_u (the number of items removed from inventory every day) is 18, but it might be as low as 16 or as high as 20. Also, our best estimate of r_i (the number of items we produce for inventory each day) is 30, but we might produce as few as 28 or as many as 32.

You should calculate the most likely inventory cost, the most pessimistic (that is, the highest the cost could be), and the most optimistic (the lowest it could be).

First, determine what combinations of variables will give the lowest and highest values. Examine the formula and note that, as r_u is in the numerator, the larger it is, the larger the value of the fraction. Further, as the fraction is being subtracted from one, the larger the fraction, the smaller the value within the parentheses. Finally, as the formula is a product of factors, the larger the expression within parentheses, the larger the overall value (inventory cost). Because r_i is in the denominator, just the opposite reasoning applies. The end result is that you find the most pessimistic result by using the lowest possible value of r_u and the highest possible value of r_i. Find the most optimistic inventory cost by using the highest possible value of r_u and the lowest possible value of r_i.

The three calculations show:

Most pessimistic $r_u = 16$ $r_i = 32$ $c_{to} = 848.13$

Most likely $r_u = 18$ $r_i = 30$ $c_{to} = 758.59$

Most optimistic $r_u = 20$ $r_i = 28$ $c_{to} = 641.13$

You are urged to take advantage of the computer programs because they make it easy for you to experiment and test the effect of any number of variables. For example, you can have the computer make complete tables of forecast calculations, each with a different smoothing factor.

Finally, each program allows you to save data you have entered into the computer's memory so you can experiment and try "what if," knowing that you can recall the original data at any time.

ABOUT THE COMPUTER PROGRAMS

The computer programs were tested on an IBM PC with a graphics board. However, the commands and functions are standard and should differ very little or not at all from the dialect another computer uses. The line most likely to be incompatible with other computers is the line (near the beginning of most of our programs) that specifies the screen mode and colors. As no special graphics are used, you can delete or modify that line in each of the programs. If you use a color monitor, you can change the color statement to get the colors that are best for you to view.

The programs are structured to call a different subroutine for each selection from the menu. Some of the subroutines call other subroutines that perform common functions. If some part of a program doesn't operate right, this structure makes it easier for you to locate the line with the typographical error.

Contents

Illustrations

Chapter 1
Key Formulas for Getting Started in
the Production Process

Chapter 2
How to Maximize the Use of Production Machines That Remove Material

Chapter 3
How to Estimate Production Time and Cost

Chapter 4
How to Measure Worker Productivity With Controlled Observation Samplings

Chapter 5
Formulas to Help Minimize Inventory Costs and Improve Ordering Procedures

Chapter 6
Quality Control and Its Impact on the Production Process

Chapter 7
Money Flow and Interest: How to Calculate Your Earning Power

Chapter 8
Calculations Related to Profit, Rent or Buy, Investment Decisions, and Depreciation

Chapter 9
Techniques of Linear Programming for Analyzing Production Choices

Chapter 10
Queuing Shortcuts: Practical Applications for Speeding up People, Machines, and Materials

Chapter 14:
Simulation: A Systematic Approach
to Testing Decisions

Appendix

CHAPTER 1

Key Formulas for Getting Started in the Production Process

This chapter introduces two fundamental aspects of the production process: the learning effect and how to calculate machine requirements for output capacities.

You'll find formulas for determining average unit time, total time needed for a given job, and a ratio based on achieving improvement goals. All these calculations are based on the principle of the learning effect, namely, that people learn parts of a new assignment rapidly at first, and the last details of an assignment at a much slower rate.

The second part of this chapter, on machine requirements, shows you how to calculate how often a machine is idle, the number of machines needed to meet a specified output, how to adjust machine requirements to allow for defects and rejects, and much more.

HOW TO MEASURE THE RATE OF PRODUCTIVITY IMPROVEMENT WITH THE LEARNING EFFECT

Individuals are subject to the learning effect and, because organizations comprise individuals, their improvement follows the same curve. Productivity increases rapidly at the start of a new task, then improvement continues, but the *rate* of improvement slows as experience is gained. This applies to all activities, including producing a product, inspecting assemblies, and checking inventory.

Symbols used

The following symbols are used in formulas for examining the learning effect.

a_1 = time to complete the task the first time

a_n = overall average unit time after completing the task n times

I = time improvement factor

k = learning factor

n = number of times the task has been completed

r_n = ratio of time to perform the task for the nth time divided by time to perform for the $(n - 1)$st time

t_n = time to perform the task the nth time

t_{tn} = total time to perform the task n times

Determining Average Unit Time

As a person (team, organization, etc.) gains experience, he or she produces each succeeding unit in less time. As the overall average time includes smaller values, it declines according to

$$a_n = \frac{a}{n^k}$$

Learning factor k is a positive constant whose value is a function of the type of task being learned. Table 1-1 shows that a relatively large k causes the overall average time to drop to a low value quickly, as would be expected with a simple task. The rate of improvement becomes negligible after the task has been completed a few times.

Smaller values of k reduce the rate of change in the beginning, but improvement continues with larger values of n. The task must be completed more often before reducing average time to any given level. A complex task will have a low k.

Example 1.1:

One position in an electronic board assembly requires that the operator insert two components, connect eleven wires, and solder nine places. The operator took ninety-six seconds the first time. What overall average can be expected after twenty boards?

Let's give k a medium value, 0.1, because the operator already knows how to wire and solder; only the new sequence has to be learned.

$$a_{20} = \frac{96}{20^{0.1}}$$

$$= 71 \text{ seconds}$$

Characteristics of the Learning Curve

According to the formula and Table 1-1, average unit time decreases continuously. This is because each time the task is completed, it takes less time than the previous completion. As average is the sum of values divided by the number of values, a decreasing number is added to the numerator and a constant (1) is added to the denominator each time the task is completed.

It should be clear that the latest time to do the task will always be less than the average unit time to that point. As n increases, the learning curve flattens and the average unit time approaches (but never equals) the latest unit time.

Knowing these characteristics of the learning curve will help you understand the formulas that follow.

Calculating the Learning Factor

Instead of assuming a value of k, you can measure improvement in time and then calculate k.

$$k = \frac{\log a_1 - \log a_n}{\log n}$$

Any base can be used for the logarithms as long as it is used consistently in the formula. Most calculators make natural logarithms more convenient by marking the ln key as a main function; common logarithms are usually a second or shifted function. The BASIC language makes natural logarithms more convenient for computer programming.

Table 1-1. Average unit time . . . drops faster when k is large.

n	Learning factor .2	.3	.4	.5	1
1	150.0	150.0	150.0	150.0	150.0
2	130.6	121.8	113.7	106.1	75.0
3	120.4	107.9	96.7	86.6	50.0
4	113.7	99.0	86.2	75.0	37.5
5	108.7	92.6	78.8	67.1	30.0
6	104.8	87.6	73.3	61.2	25.0
7	101.6	83.7	68.9	56.7	21.4
8	99.0	80.4	65.3	53.0	18.8
9	96.7	77.6	62.3	50.0	16.7
10	94.6	75.2	59.7	47.4	15.0
11	92.9	73.1	57.5	45.2	13.6
12	91.3	71.2	55.5	43.3	12.5
13	89.8	69.5	53.8	41.6	11.5
14	88.5	68.0	52.2	40.1	10.7
15	87.3	66.6	50.8	38.7	10.0
16	86.2	65.3	49.5	37.5	9.4
17	85.1	64.1	48.3	36.4	8.8
18	84.1	63.0	47.2	35.4	8.3
19	83.2	62.0	46.2	34.4	7.9
20	82.4	61.1	45.3	33.5	7.5
21	81.6	60.2	44.4	32.7	7.1
22	80.8	59.3	43.6	32.0	6.8
23	80.1	58.6	42.8	31.3	6.5
24	79.4	57.8	42.1	30.6	6.3
25	78.8	57.1	41.4	30.0	6.0
26	78.2	56.4	40.7	29.4	5.8
27	77.6	55.8	40.1	28.9	5.6
28	77.0	55.2	39.6	28.3	5.4
29	76.5	54.6	39.0	27.9	5.2
30	76.0	54.1	38.5	27.4	5.0

Example 1.2

Several new employees were timed as they learned to pack a delicate assembly in its foam supports, insert it into a box, include an instruction book and five small items, and seal the box. Their average the first time was 4.12 minutes. The overall average after 50 packings was 3.37 minutes. What overall average can you expect after 150 packings?

To get the answer, you should apply the same formula to the average results from observing several individuals. You must first solve for k and then use that value in the first formula to find overall average.

$$k = \frac{\log 4.12 - \log 3.37}{\log 50}$$

$$= .0514$$

Now use $k = .05$ in the first formula

$$a_{150} = \frac{4.12}{150^{.05}}$$

$$= 3.21 \text{ minutes}$$

Figuring Total Time for a Job

Total time to complete a task n times, given k and the original time, is

$$t_{tn} = t_1 n^{1-k}$$

Example 1.3

In example 1.1, how much time do you expect the operator to spend (cumulative time) on the first twenty boards?

$$t_{20} = 96(20)^{1-0.1}$$

$$= 1,423 \text{ seconds}$$

$$= 23.72 \text{ minutes}$$

Determining Final Time for a Given Task

The time required to do the task the nth time is found from

$$t_n = a_1 \left(\frac{1 - k}{n^k} \right)$$

Example 1.4

In example 1.2 you saw that workers should average 3.21 minutes apiece for the first 150 assemblies packed. How long should the 150th one take?

$$t_{150} = 4.12 \left(\frac{1 - 0.05}{150^{.05}} \right)$$

$$= 3.05 \text{ minutes}$$

The fact that this individual time is close to the average time suggests that the learning curve is flattening, and changing slowly. You can check by finding how long the 151st packing should take.

$$t_{151} = 4.12 \left(\frac{1 - 0.05}{151^{.05}} \right)$$

$$= 3.05 \text{ minutes}$$

As both of these times, rounded to two decimal places, are the same, it is safe to say that the learning curve has flattened out.

Achieving Your Improvement Goal:
The Trial-to-Trial Ratio

Common sense tells us that improvement does not continue to the point where the task can be done in zero time. One good approach is to say that the curve is flat after the ratio of times between successive trials reaches a given value.

What should that ratio be? As task completion time decreases with each trial, a ratio of any completion time divided by the preceding completion time will be larger than zero and less than one. Ratios such as 0.99, 0.995, and 0.999 are often used.

The improvement ratio

The following formula finds how many times a task must be completed before improvement flattens to a given ratio.

$$n \ge \frac{1}{1 - r^{1/k}}$$

The inequality sign means that, where there is a decimal part to the answer, use the next largest integer.

Example 1.5

New order clerks require one hour to locate all the information, complete an order form, and get signatures. The learning factor is .08. A clerk is said to know the job where there is less than one-half of one percent improvement in time from one order to the next. That is, the ratio of times between succeeding orders will be 0.995 or larger. How long will it take an experienced clerk to process an order?

You first determine how many orders must be processed before you can expect the ratio of times to be 0.995. Then you will use that value of n to calculate processing time.

$$n \ge \frac{1}{1 - 0.995^{1/.08}}$$

$$\ge 16.465$$

Because this answer has a decimal part, you use the next largest integer, 17. Next, determine how long it should take to process an order for the seventeenth time.

$$t_{17} = 60 \left(\frac{1 - 0.08}{17^{.08}} \right)$$

$$= 44 \text{ minutes}$$

Because this calculation involves fractions and exponents, it is easier to find n from Table 1-2. Simply locate the intersection of appropriate values of r and k.

All entries have been adjusted to give the next largest integer when the calculation shows a decimal part.

How to Find a Reasonable Goal for Improvement in Unit Time

The next formula finds how many times the tasks should be performed before there will be a given improvement in unit time.

$$n_2 = \frac{n_1}{I^{1/k}}$$

This formula by itself may yield ludicrous values, such as doing the task billions of times. The interpretation is that you are asking for an improvement that cannot realistically be achieved. You can use the preceding formula to find a range of reasonable improvement expectations.

Example 1.6

One typical team, with a learning factor of 0.09, took 40 minutes to complete a task for the 25th time. After how many times can you expect them to do the task in 20 minutes?

$$n_2 = \frac{25}{0.5^{1/.09}}$$

$$= 55,299$$

Table 1-2. Number of trials required to bring ratio of successive times down to value shown.

Ratio	.01	.02	.03	0.5	.08	.1	.2	.3	.5	.8	1	2	3	5
0.900	2	2	2	2	2	2	3	4	6	9	10	20	29	48
0.910	2	2	2	2	2	2	3	4	6	9	12	22	33	54
0.920	2	2	2	2	2	2	3	5	7	11	13	25	37	61
0.930	2	2	2	2	2	2	4	5	8	12	15	29	42	70
0.940	2	2	2	2	2	3	4	6	9	14	17	33	49	82
0.950	2	2	2	2	3	3	5	7	11	17	20	40	59	98
0.960	2	2	2	2	3	3	6	8	13	21	25	50	74	123
0.970	2	2	2	3	4	4	8	11	17	27	34	67	99	165
0.980	2	2	3	4	5	6	11	16	26	41	50	100	149	248
0.990	2	3	4	6	9	11	21	31	51	81	100	200	299	498
0.991	2	3	4	7	10	12	23	34	56	89	112	222	333	554
0.992	2	4	5	7	11	13	26	38	63	101	125	250	374	623
0.993	2	4	5	8	12	15	29	44	72	115	143	286	428	713
0.994	3	4	6	9	14	18	34	51	84	134	167	333	499	832
0.995	3	5	7	11	17	21	41	61	101	161	200	400	599	998
0.996	4	6	8	13	21	26	51	76	126	201	250	500	749	1248
0.997	4	8	11	18	28	34	68	101	167	267	334	667	999	1665
0.998	6	11	16	26	41	51	101	151	251	401	500	1000	1499	2498
0.999	11	21	31	51	81	101	201	301	501	800	1000	2000	3000	4998

It is not reasonable that improvement would continue on the same curve for that long. To continue the investigation, let us find how many times they should do the task to bring their learning curve to within one tenth of one percent of a straight line. That is, to make the ratio of times between succeeding completions to 0.999.

$$n \geq \frac{1}{1 - 0.999^{1/.09}}$$

$$\geq 90.456$$

After the team does the task ninety-one times, we cannot expect any significant further improvement. It is not likely they will ever get their time down to twenty minutes.

COMPUTER PROGRAM FOR SOLVING THE LEARNING FUNCTION FORMULAS

Figure 1-1 shows a computer program that solves the formulas given in this section. Although it was written specifically for an IBM PC the commands are standard and should work with little or no modification on any computer that uses BASIC.

The line most likely to require changing for other computers is line 60. CLS is a command to clear the screen; some computers may respond to commands such as HOME, CALL CLEAR, or others.

A good way to test the program after you key it into your computer is to enter the numbers from the examples in this section. The program is structured with a subroutine for each item in the menu, so if any result is different from the example, you can easily find the area of the program that does the calculation. Then carefully proofread it against the listing in the book.

MACHINE REQUIREMENTS: BASIC FORMULAS FOR FINDING OUTPUT CAPACITIES

Formulas in the next series find ratios and numbers that apply to machines and their output.

Symbols used

The following symbols are used in this section:

e_c = efficiency of capital equipment, percent of running time

f_a = fraction of output accepted

   = U_a/U

f_r = fraction of output rejected

   = U_r/U

Figure 1-1.   Computer program for learning function calculations.

```
10 '************************************************************
20 '** LRN.BAS  Calculations based on learning function    **
30 '** equations.  Written for Handbook of Manufacturing   **
40 '** and Production Management Formulas, Charts, and      **
50 '** Tables                                               **
60 '************************************************************
70 CLEAR,,,32768!: SCREEN 6: COLOR 3,2
80 CLS
90 GOSUB 1010                     'Display menu
100 CLS
110 ON SELECT GOSUB 1210,1410,1610,1810
120 GOTO 80
130 END
1000 'Subroutine to display menu  <<<<<<<<<<<<<<<<<<<<
1010 PRINT
1020 PRINT "1  Calculate times -- average, total, final"
1030 PRINT "2  Calculate learning factor"
1040 PRINT "3  Goal based on ratio of succeeding times"
1050 PRINT "4  Goal based in improvement ratio"
1060 PRINT: INPUT "Select by number ",SELECT
1070 IF SELECT>0 AND SELECT<5 THEN RETURN
1080 PRINT "You can select only between 1 and 4"
1090 GOTO 1010
1200 '(1) Subroutine to Calculate times  111111111111111
1210 GOSUB 2010                    'Ask for t1 and n
1220 INPUT "Enter value for k ",K
1230 AN=T1/N^K
1240 TT=T1*N^(1-K)
1250 FT=T1*((1-K)/N^K)
1260 PRINT: PRINT "The average time to do the task"
1270 PRINT N;"times is";AN;".  It will take a"
1280 PRINT "total time of";TT;"and the last trial"
1290 PRINT "will take";FT;"time units."
1300 PRINT
1310 INPUT "Press <Enter> to return to main menu ",E$
1320 RETURN
1400 '(2) Subroutine to calculate learning factor  222222222222222
1410 GOSUB 2010                    'Ask for t1 and n
1420 INPUT "Enter average unit time ",AN
1430 IF AN>0 AND AN<T1 THEN 1480
1440 PRINT "Average unit time must be positive and less"
1450 PRINT "than the time for the first trial.  Please answer"
1460 PRINT "the question again."
1470 PRINT: GOTO 330
1480 K=(LOG(T1)-LOG(AN))/LOG(N)
1490 PRINT: PRINT "When the first trial takes";T1;"time units"
1500 PRINT "and the average unit time for";N;"trials is"
1510 PRINT AN;", the learning factor is";K
1520 PRINT
1530 INPUT "Press <Enter> to return to main menu ",E$
1540 RETURN
1600 '(3) Subroutine for Goal based on ratio of succeeding times  33333333
1610 INPUT "Enter learning factor ",K
1620 INPUT "Enter goal ratio ",R
1630 N=1/(1-R^(1/K))
1640 IF N=INT(N) THEN 1660
1650 N=INT(N)+1
1660 PRINT: PRINT "When the learning factor is";K;", the ratio"
1670 PRINT "between successive times will be";R;"or more"
1680 PRINT "for all trials beginning with number";N
1690 PRINT
1700 INPUT "Press <Enter> to return to main menu ",E$
1710 RETURN
```

Figure 1-1. (continued)

```
1800 '(4) Subroutine for Goal based on improvement ratio  4444444444444444
1810 INPUT "Enter learning factor ",K
1820 PRINT "You are looking for a trial whose time is what part"
1830 INPUT "of the present trial's time? ",I
1840 IF I>0 AND I<1 THEN 1890
1850 PRINT "The improvement factor must be a positive decimal"
1860 PRINT "larger than 0 and less than 1 (such as .8).  Please"
1870 PRINT "answer the question again."
1880 PRINT: GOTO 1820
1890 INPUT "What is the number of the present trial? ",N1
1900 N2=N1/(I^(1/K))
1910 PRINT: PRINT "When the learning factor is";K;", trial number"
1920 PRINT N2;"will take";I;"as long as trial number";N1
1930 PRINT
1940 INPUT "Press <Enter> to return to main menu ",E$
1950 RETURN
2000 'Subroutine to ask for t1 and n  <<<<<<<<<<<<<<<<<
2010 INPUT "Enter time to complete task the first time ",T1
2020 INPUT "How many trials are you working with? ",N
2030 RETURN
```

n = number of machines

t = time to manufacture one unit, minutes

T_a = time per shift that machine actually produces, minutes

T_t = hours worked per day

   = 8 times number of shifts

T_y = production hours per year, usually 2080

U = units per shift

   = $U_a + U_r$

U_a = units accepted per shift

U_r = units rejected per shift

U_y = units manufactured per year

Machine Utilization Ratio: Determining How Often a Machine Is Idle

Machine downtime can be unplanned, such as for breakdowns, or it can be planned, such as for preventive maintenance. Idle time can also be scheduled. Therefore, the portion used of available machine time is a figure basic to analyzing production.

$$e_c = \frac{\text{time the machine is working}}{\text{time the machine could be working}}$$

A ratio of 1 means there is no idle time. Some analysts prefer to multiply the fraction by 100, giving *percent* utilization.

For some investigations it is more meaningful to calculate the portion of time a machine is *idle*. Then the numerator should be idle time instead of working time.

Example 1.7

When an electronic test station is turned on in the morning, a technician runs checks on it for fifteen minutes before certifying it for the day. It is then available for the rest of the eight-hour shift. During a typical shift, 120 assemblies are tested, each using three and a half minutes on the tester, including hookup time. What is the utilization ratio of the tester? We will multiply the hour figures by sixty to keep all quantities in minutes.

$$e_c = \frac{120(3.5)}{7.75(60)}$$

$$= 0.9032$$

The machine is used just over ninety percent of its available time. If the emphasis is on idle time, then it is idle just under ten percent of the time.

How to Calculate the Number of Machines Needed to Meet Output

When total output is specified, the number of machines required can be calculated from this formula. It has been adjusted for a mixture of units (unit manufacturing time in minutes; length of workday in hours).

$$n = \frac{1.67tU}{T_t e_c}$$

Example 1.8

You want to produce 2,000 units per shift on machines that each complete one unit every four minutes. The production department works one eight-hour shift with two ten minute breaks; the machines run seven hours and forty minutes per day. How many machines will be required?

As efficiency is given in percent, you must first convert running time to percentage of shift.

$$e_c = \left(\frac{7 \text{ hours, 40 minutes}}{8 \text{ hours}} \right) 100$$

$$= 95.83 \text{ percent}$$

These figures can be substituted directly into the formula for n.

$$n = \frac{1.67(4)2,000}{8(95.83)}$$

$$= 17.43$$

To decide how to round the forty-three-hundredths of a machine, question the goal of 2,000 per day. Do you want exactly 2,000 units? If so, you will need eighteen machines, with running time of less than seven hours and forty minutes (see the next set of formulas). If 2,000 is just a run-it-up-the-flagpole number, you make a decision after determining how many units you can make with seventeen machines and how many with eighteen machines (see the next set of formulas).

Alternative Formulas for Finding Number of Machines Required

Each of the following formulas is a rearrangement of the preceding formula — each arrangement is solved for finding a different quantity.

Formulas in the first column use e_c percentage efficiency of time usage. Those in the second column use actual running time instead.

Percentage efficiency   Actual running time

$$t = \frac{0.6\, n\, T_t e_c}{U} \qquad\qquad\qquad t = \frac{60\, n t_a}{U}$$

$$U = \frac{0.6\, n\, T_t e_c}{t} \qquad\qquad\qquad U = \frac{60\, n t_a}{t}$$

$$T_t = \frac{1.67\, t U}{n\, e_c} \qquad\qquad\qquad T_a = \frac{0.0167\, t U}{n}$$

$$e_c = \frac{1.67\, t U}{n\, T_t}$$

Example 1.9

How many hours a day must you operate six machines that each produce one unit every fifty seconds if you need 4,000 units per day?

To find hours per day (T_a), substitute directly into the last equation in the right-hand column. Be careful with t; it is defined in minutes but given in seconds. Fifty seconds is $\frac{50}{60}$ of a minute.

Hours per day

$$T_a = \frac{0.0167(50/60)4{,}000}{6}$$

$$= 9.28 \text{ hours}$$

The facts of the example indicate that adding more machines is not one of your choices, so you must either extend the shift to about $9\frac{1}{2}$ hours or do part of the production on another shift.

Example 1.10

Example 1.8 mentioned calculating total production from seventeen and from eighteen machines. You can now do that with the second formula in the left-hand column.

Units per shift

$$U = \frac{0.6(17)8(95.83)}{4}$$

$$= 1{,}955 \text{ units from } 17 \text{ machines}$$

$$U = \frac{0.6(18)8(95.83)}{4}$$

$$= 2{,}070 \text{ units from } 18 \text{ machines}$$

Adjusting Output to Allow for Defects

If a certain fraction of output is usually rejected, the following formula shows how many to produce in order to have a given number of acceptable units.

Acceptable units

$$U = \frac{U_a}{1 - f_r}$$

$$= \frac{U_a}{f_a}$$

Example 1.11

A production department normally rejects three percent of a machine's output. How many pieces should they produce in order to have 1,275 acceptable ones?

Three percent means three per hundred ($\frac{3}{100}$), so f_r is 0.03.

$$U = \frac{1{,}275}{1 - .03}$$

$$= 1{,}315$$

This formula assumes that all rejects are scrapped. If it is financially efficient to repair the rejects, then 1,275 must be started if 1,275 are to be shipped.

Adjusting Machine Requirements to Allow for Rejects

This formula determines how many machines are required for a given amount of shippable output, considering that a given portion of production is rejected.

$$n = \frac{U_y\, t}{0.6 e_c T_y\, (1 - f_r)}$$

Example 1.12

A manufacturer of bowling pins must have 255,000 pins ready to ship each year. One out of 230 generally has a grain pattern that might split, so it is destroyed. The turning machines, each of which makes forty-two pins per hour, work eighty-nine percent of the time and the factory operates on a normal 2080-hour year. How many turning machines are required?

Because t is defined in minutes, we will use 60/42; it takes sixty minutes to make forty-two pieces.

$$n = \frac{255,000(60/42)}{0.6(89)2080(1 - 1/230)}$$

$$= 3.29$$

The firm will need four machines, but that will change the utilization ratio. To calculate the new ratio, rearrange the formula to solve for e

$$e = \frac{U_y t}{0.6 n T_y (1 - f_r)}$$

$$= \frac{255,000(60/42)}{0.6(4)2080(1 - 1/230)}$$

$$= 73.29 \text{ percent}$$

It was expected that the machines would run eighty-nine percent of the available time, but that led to a requirement that included a fraction of a machine. When you use a whole number of machines, they will run seventy-three percent of the time.

Sequence of Machines: Determining How Many You Need to Meet Output

Usually the output of one set of machines will move on to other sets of machines. Our goal here is to determine how many items should be started at the first set of machines so that a given number will be produced for shipping. For each stage,

$$\text{Number needed at start} = \frac{\text{Output needed}}{1 - f_r}$$

Using this formula for each stage is best explained with an example.

Example 1.13

The first operation grinds a casting, and one percent of the machined items are rejected instead of being passed on to the next operation. Buffing and polishing is second; 0.5 percent of the output from this set of machines is rejected. Problems with the third set of machines (nameplate attachment) result in a rejection rate of two percent. How many units should be started into the sequence if we are to ship 30,000, and all rejects are scrapped?

Apply the formula to each stage, starting at the last one.

$$\text{Start}_3 = \frac{30{,}000}{1 - 0.02}$$

$$= 30{,}613$$

Then determine how many units should be started at the second stage in order to have 30,613 units at the input to the third stage.

$$\text{Start}_2 = \frac{30{,}613}{1 - .005}$$

$$= 30{,}767$$

Finally, determine how many units should be started at the first stage.

$$\text{Start}_1 = \frac{30{,}767}{1 - 0.01}$$

$$= 31{,}078$$

Therefore, at least 31,078 units should be started at the first stage of this manufacturing process.

CHAPTER 2

How to Maximize the Use of Production Machines That Remove Material

This chapter looks at machines that remove material, not machines that bend material, stamp material, shear material, and so forth. After those machines have done their jobs, the material weighs just as much as it did before; the machines do not produce scrap. This chapter looks at machines that *do* remove material, so their output is the machined product plus scrap. You'll find formulas for determining the power required, the feed rate, and the volume of material removed. There are separate sections on cutting tools, drills that cut by rotating, how to measure the speed and rotation of machines, and a computer program to help you solve the formulas given throughout the chapter.

CUTTING TOOLS: HOW TO FIND THE REQUIRED FORCE FOR OPERATING

This section gives formulas for finding forces required by milling machines, drills, and other tools that cut or cut into material. Machine setting for minimum cost is also included.

Symbols used

The following symbols are used in this section.

a = cross-sectional area of chip or piece, square inches
  = depth of cut times feed (one spindle revolution)
c_l = cost of rejecting piece outside lower limit
c_u = cost of rejecting piece outside upper limit
d = diameter of work or tool
g = goal for machine setting, inches
k = constant, determined from tables by the formulas
L_l = lower limit for accepting piece
L_u = upper limit for accepting piece
p = pressure at point of cut, pounds
s = cutting speed, feet per minute
T = tolerance, inches
w = horsepower at tool
σ = standard deviation

Table 2-1. Constant for pressure and horsepower formulas.

Type of material	Constant k
Bronze	1.26
Cast iron	1.69
Cast steel	2.93
Mild Steel	4.12
High carbon steel	5.06

Calculating Pressure

A cutting tool is subjected to pressure at the point of cut, as calculated from the following;

$$p = 80{,}300(1.33)^k a$$

where k is a constant, given in Table 2-1.

Example 2.1

A cylinder of cast iron is being cut by a lathe to a depth of .06 inch, with a feed of .0286 inch per revolution. How much pressure is at the cut?

$$p = 80{,}300(1.33)^{1.69} (.06) (.0286)$$
$$= 223 \text{ pounds}$$

Finding Required Horsepower

A formula similar to the formula for pressure finds the horsepower required at the tool's cutting edge.

$$w = 2{,}433(1.33)^k a s$$

where k is a constant given in Table 2-1.

Example 2.2

A tool cuts mild steel to a depth of .045 inch, feeding at .020 inch. What horsepower is required at the cutting edge if the cutting speed is 70 feet per minute?

$$w = 2.433(1.33)^{4.12} (.045) (.020) (70)$$

$$= 0.4963 \text{ horsepower}$$

Machine Accuracy

The dimensional tolerance to which a piece can be held is a function of both the machining operation and the piece's dimension.

$$T = kd^{0.371}$$

Table 2-2.   Constant k for machining accuracy formula.

Machining operation	k
Drilling, rough turning	.007816
Finish turning, milling	.004812
Turning on turret lathe	.003182
Automatic turning	.002417
Broaching	.001667
Reaming	.001642
Precision turning	.001378
Machine grinding	.001008
Honing	.000684

Coefficient k is a function of the machining operation, as shown in Table 2.2.

Example 2.3

A bar with 0.875-inch diameter is being turned on a turret lathe. What is a reasonable tolerance?

$$T = .003182(0.875)^{0.371}$$

$$= .003028 \text{ inch}$$

Calculating the Goal Dimension

This formula finds nominal measurement for machining, which may not be the midpoint of the tolerance band. This setting will minimize costs under the following conditions:

a. If the part has not been machined enough, it can be reworked.
b. If the part has been machined too far, it must be scrapped.

Work coming from an automatic machine can be expected to have a normal distribution about the center value to which the machine is set. However, it is not always desirable for the distribution to be centered about the design value because the cost of correcting rejects may not be the same on each side. This formula applies when it costs more to reject a piece because of too-small dimensions than because of too-large dimensions.

It will therefore tell us to set the machine for a goal dimension on the large side of center, so that most of the rejects will be due to too-large dimensions.

$$g = \frac{1}{2(L_u - L_l)}\left[L_u{}^2 - L_l{}^2 + \sigma^2 \, ln\left(\frac{c_l}{c_u}\right)\right]$$

Example 2.4

A solid cylinder is to be turned down to 1.125 inches diameter by a machine under robotic control. If a piece comes off the machine measuring 1.128 or larger, a new shop order must be cut to have the piece machined further. That procedure costs an additional $10.80.

When a piece measures 1.122 or smaller, it must be scrapped at a total cost of $23.35. Standard deviation of the machined measurements is .002 inch. To what dimension should the machine be set?

$$g = \frac{1}{2(1.128 - 1.122)}\left[1.128^2 - 1.122^2 + .002^2 \, ln \, \frac{23.35}{10.80}\right]$$

$$= 1.1253 \text{ inches}$$

As it costs more to scrap a piece than to rework one, this result, telling us to aim a little on the high side, seems reasonable.

Table 2-3. Results of computer simulation.

Setting	Cost
1.1240	19,636.40
1.1241	19,302.51
1.1242	16,651.72
1.1243	15,875.94
1.1244	15,675.74
1.1245	14,460.86
1.1246	13,253.52
1.1247	13,087.17
1.1248	12,534.68
1.1249	12,098.63
1.1250	11,515.49
1.1251	10,905.50
1.1252	11,179.89
1.1253	9,823.86
1.1254	10,904.98
1.1255	10,231.94
1.1256	10,055.39
1.1257	10,773.72
1.1258	10,495.58
1.1259	10,698.77
1.1260	11,627.85

Using Computer Simulation to Determine Optimum Setting

Another way to examine the question of optimum setting is with a computer simulation. Chapter 14 describes several simulations and uses the previous situation as one of its examples.

One advantage of the simulation over the formula is that the results of other settings can be compared. In addition to showing that a setting of 1.1253 results in the lowest cost, the simulation can show likely costs for that or any other setting.

Table 2-3 shows the results of simulating 5,000 pieces at each .0001 increment from 1.1240 to 1.1260.

A simulation, like real life, includes an element of randomness, so identical results are not expected every time. The randomness also explains why the cost column in Table 2-3 does not show smooth changes. In fact, there are some reversals from 1.1251 through 1.1259. Simulating 10,000 units instead of 5,000 would probably show fewer reversals. An extremely large simulation would be expected to show a more orderly progression in costs.

HOW TO APPLY FORMULAS TO DRILL FORCES

Formulas in the next series apply to drills that cut by rotating, and do not include other types such as star drills.

Symbols used

The following symbols are used in this section.

d = diameter of drill, inches

f = feed per revolution, inches

f_{hp} = horsepower equivalent of torque

f_{th} = thrust, pounds

f_{tq} = torque, inch-pounds

v = rotational velocity, rotations per minute (rpm)

Drill Thrust

The amount of thrust required for drilling is found from the following formula.

$$f_{th} = 57.5f^{0.8}d^{1.8} + 625d^2$$

Example 2.5

A $\frac{3}{8}$ inch drill is feeding at .004 inch per revolution. What is the thrust?

$$f_{th} = 57.5(.004)^{0.8}(0.375)^{1.8} + 625(0.375)^2$$

$$= 88.01 \text{ pounds}$$

Drill Torque

This formula calculates the torque for a drilling operation.

$$f_{tq} = 25.2f^{0.8}d^{1.8}$$

Example 2.6

Calculate the torque for the preceding example.

$$f_{tq} = 25.2(.004)^{0.8}(0.375)^{1.8}$$

$$= 0.052 \text{ inch-pound}$$

Drill Horsepower

This formula finds the horsepower equivalent of a torque.

$$f_{hp} = 15.87 \times 10^{-6}f_{tq}v$$

$$= 400 \times 10^{-6}f^{.8}d^{1.8}v$$

Example 2.7

What is the equivalent horsepower in Example 2.6 if the drill rotates at 650 rpm?

$$f_{hp} = 15.87 \times 10^{-6}(0.052)650$$

$$= 0.0005 \text{ horsepower}$$

HOW TO CALCULATE THE SPEED AND ROTATION OF MACHINES

This section looks at turning speed, surface speed, and rotational speed of tools and work.

Symbols used

The following symbols are used in this section. Dimensions given are used in the formulas.

a_l = linear feed, or advance, in inches per minute

a_r = feed, in inches per revolution

a_t = feed, in inches per tooth

d = diameter of work where tool is operating, inches

r = radius of work where tool is operating, inches

s = cutting speed, feet per minute

s_r = rotational speed of work, revolutions per minute

s_s = surface speed of work, feet per minute

Shaft Speed

Cylindrical shapes passing a contact point with a tool, as in a lathe, should move at a speed recommended in the manufacturer's literature. This formula translates surface speed of the cylinder to revolutions per minute.

$$s_r = \frac{12s_s}{\pi d}$$

$$= \frac{6s_s}{\pi r}$$

Example 2.8

A solid metal cylinder with $1\frac{1}{4}$ inch diameter is to be worked in a lathe. Its recommended surface speed is 240 feet per minute. How fast should the lathe turn?

$$s_r = \frac{12(240)}{3.1416(1.25)}$$

$$= 733 \text{ revolutions per minute}$$

The nearest available speed to 733 rpm will bring the surface speed to its recommended value.

Example 2.9

A lathe cuts at a speed of 102 feet per minute on cylindrical work with a diameter of 4.385 inches. What is the spindle speed?

$$s_r = \frac{12(102)}{3.1416(4.385)}$$

$$= 88.85 \text{ revolutions per minute}$$

Example 2.10

The material should not be cut faster than 200 feet per minute. What is the fastest rotation allowed when the material's radius is $2\frac{1}{2}$ inches?

$$s_r = \frac{1.9099(200)}{2.25}$$

$$= 152.79 \text{ revolutions per minute.}$$

Rotational speed should be kept below 153 feet per minute.

Surface Speed

This formula finds the rate at which the work's surface passes the tool.

$$s_s = 0.2618ds_r$$

$$= 0.5236rs_r$$

Example 2.11

Cylindrical material of 1.12 inch diameter rotates at 460 rpm. What is the work's surface speed?

$$s_s = 0.2618(1.12)460$$

$$= 134.88 \text{ feet per minute}$$

Tool Feed per Revolution

The rate at which the tool advances into or past the work is given by this formula.

$$a_r = \frac{a_l}{s_r}$$

$$= \frac{0.2618\, d\, a_l}{s_s}$$

$$= \frac{0.5236\, r\, a_l}{s_s}$$

Example 2.12

The tool is to advance forty inches per minute when the work has a diameter of 0.875 inch. If the surface speed is 180 feet per minute, at what rate should the tool advance?

$$a_r = \frac{0.2618(0.875)40}{180}$$

$$= 0.0509 \text{ inch per revolution}$$

Tool Feed per Minute

This formula is a rearrangement of Example 2.12; it is used when the rotational speed is known.

$$a_l = a_r s_r$$

$$= \frac{3.8197 a_r s_s}{d}$$

$$= \frac{1.9099 a_r s_s}{r}$$

Example 2.13

The tool advances 0.075 inch each revolution and the work is rotating 660 revolutions per minute. What is the feed rate?

$$a_l = 0.075(660)$$

$$= 49.5 \text{ inches per minute}$$

Tool Feed per Tooth

This formula is used when the cutter information is known in terms of teeth instead of time.

$$a_t = \frac{a_r}{t}$$

$$= \frac{a_l}{t s_r}$$

Example 2.14

A cutter with 33 teeth advances 0.06 inch per revolution. At what rate does it advance?

$$a_t = \frac{0.06}{33}$$

$$= 0.0018 \text{ inch per tooth}$$

HOW TO FIND THE VOLUME OF MATERIAL REMOVED

Formulas in this section find the volume of material removed by operations such as cutting, drilling, and milling. In many applications, the results of these formulas (volume) will be multiplied by unit weight of the material to find the weight of material removed.

Symbols used

The following symbols are used in this section.

d = diameter of hole or tool, inches

e = distance cut extends on surface, inches

f = linear feed rate, inches per minute

h = depth of cut, inches

t = angle of tip of drill, degrees

v = volume, cubic inches

v_r = rate of volume removal, cubic inches per minute

w = width of cut, inches

Time Rate for Material Removed in a Rectangular Cut

This formula gives the rate at which material is removed when the cut has a rectangular shape.

$$v_r = whf$$

Example 2.15

A milling machine is cutting a $\frac{1}{8}$-inch deep groove with the face of a $\frac{5}{8}$-inch diameter tool. It feeds at 28 inches per minute. At what rate is material removed?

$$v_r = 0.625(0.125)28$$

$$= 2.1875 \text{ cubic inches per minute}$$

Linear Feed for Rectangular Material Removal

This formula is a rearrangement of Example 2.15; it is used to find the feed rate required for a given rate of material removal.

$$f = \frac{v_r}{wh}$$

Example 2.16

You want to remove three cubic inches per minute while milling $\frac{1}{8}$ inch from the side of a block that is $1\frac{1}{16}$ inches thick. How fast should the cutter feed if you are cutting with the circumference of a tool at least $1\frac{1}{16}$ inches long?

$$f = \frac{3}{1.0625(0.125)}$$

$$= 22.6 \text{ inches per minute}$$

Material Removed by Drill

This formula applies when material is removed by a drill. One version of the formula approximates the volume by assuming the drill is a plain cylinder without a triangular tip. The second version is precise and includes a correction for the angle of the drill's tip.

$$v = \frac{\pi d^2 h}{4}$$

This determines the volume of a hole that is completely cylindrical. It is correct when the drill goes all the way through the material, and is close when the drill stops inside the material. If the drill goes through, then h is the thickness of the material.

The complete formula, considering that the drill's tip is triangular, is

$$v = \frac{\pi d^2}{4}\left[h - \frac{d}{3}\cot\frac{t}{2}\right]$$

Example 2.17

A $\frac{3}{4}$-inch hole is drilled one inch into thick material, using a twist drill with 118-degree tip. How much material is removed?

$$v = \frac{\pi 0.75^2}{4}\left[1 - \frac{0.75}{3}\cot\frac{118°}{3}\right]$$

$$= 0.3754 \text{ cubic inch}$$

If you had used the simplified formula, it would have shown 0.4418 cubic inch, or about eighteen percent too high. The difference between the formulas diminishes with smaller diameter drills, and with deeper holes.

Figure 2-1.   Slot, with cutter overtravel.

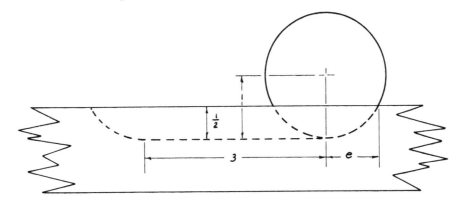

Table 2-4. Cutter travel.

Depth of cut	1	2	3	4	Radius of cutter 5	6	7	8	9	10
0.10	0.436	0.624	0.768	0.889	0.995	1.091	1.179	1.261	1.338	1.411
0.15	0.527	0.760	0.937	1.085	1.216	1.333	1.441	1.542	1.636	1.726
0.20	0.600	0.872	1.077	1.249	1.400	1.536	1.661	1.778	1.887	1.990
0.25	0.661	0.968	1.199	1.392	1.561	1.714	1.854	1.984	2.107	2.222
0.30	0.714	1.054	1.308	1.520	1.706	1.873	2.027	2.170	2.304	2.431
0.35	0.760	1.130	1.406	1.636	1.838	2.019	2.186	2.340	2.485	2.622
0.40	0.800	1.200	1.497	1.744	1.960	2.154	2.332	2.498	2.653	2.800
0.45	0.835	1.264	1.580	1.843	2.073	2.280	2.469	2.645	2.810	2.966
0.50	0.866	1.323	1.658	1.936	2.179	2.398	2.598	2.784	2.958	3.122
0.55	0.893	1.377	1.731	2.024	2.280	2.509	2.720	2.915	3.098	3.271
0.60	0.917	1.428	1.800	2.107	2.375	2.615	2.835	3.040	3.231	3.412
0.65	0.937	1.476	1.865	2.186	2.465	2.716	2.946	3.159	3.358	3.546
0.70	0.954	1.520	1.926	2.261	2.551	2.812	3.051	3.273	3.480	3.676
0.75	0.968	1.561	1.984	2.332	2.634	2.905	3.152	3.382	3.597	3.800
0.80	0.980	1.600	2.040	2.400	2.713	2.993	3.250	3.487	3.709	3.919
0.85	0.989	1.636	2.092	2.465	2.789	3.079	3.343	3.589	3.818	4.035
0.90	0.995	1.670	2.142	2.528	2.862	3.161	3.434	3.686	3.923	4.146
0.95	0.999	1.702	2.190	2.588	2.932	3.240	3.521	3.781	4.025	4.254
1.00	1.000	1.732	2.236	2.646	3.000	3.317	3.606	3.873	4.123	4.359
1.50		1.936	2.598	3.122	3.571	3.969	4.330	4.664	4.975	5.268
2.00		2.000	2.828	3.464	4.000	4.472	4.899	5.292	5.657	6.000
2.50			2.958	3.708	4.330	4.873	5.362	5.809	6.225	6.614
3.00			3.000	3.873	4.583	5.196	5.745	6.245	6.708	7.141
3.50				3.969	4.770	5.454	6.062	6.614	7.124	7.599
4.00				4.000	4.899	5.657	6.325	6.928	7.483	8.000
4.50					4.975	5.809	6.538	7.194	7.794	8.352
5.00					5.000	5.916	6.708	7.416	8.062	8.660
5.50						5.979	6.837	7.599	8.292	8.930
6.00						6.000	6.928	7.746	8.485	9.165
6.50							6.982	7.858	8.646	9.367
7.00							7.000	7.937	8.775	9.539
7.50								7.984	8.874	9.682
8.00								8.000	8.944	9.798
8.50									8.986	9.887
9.00									9.000	9.950
9.50										9.987
10.00										10.000

HOW TO DETERMINE DEPTH OF PARTIAL CUT

When a circular cutter such as a milling machine or a saw blade does not part the material, we are interested in the depth of cut. This formula determines the cutter's travel from the point of first contact to the point where it cuts full

Figure 2-2. Computer program for formulas in Chapter 2.

```
10 '****************************************************************
20 '** CHAP2.BAS  Program to solve formulas in Chapter 2 of   **
30 '** Handbook of Manufacturing and Production Management    **
40 '** Formulas, Charts, and Tables                           **
50 '****************************************************************
60 CLEAR,,,32768!: SCREEN 6: COLOR 3,3
70 CLS
80 GOSUB 1010                     'Display menu
90 CLS
100 ON SELECT GOSUB 1410,1610,1810,2010,2410,2610,2810,3110,3310,3510,3710,4010,
4310,4610,4810,5010,5210
110 GOTO 70
120 END
1000 'Subroutine to display menu  <<<<<<<<<<<<<<
1010 PRINT TAB(34)"CUTTING TOOLS"
1020 PRINT " 1 Tool pressure"; TAB(40)" 3 MACHINING ACCURACY"
1030 PRINT " 2 Horsepower"; TAB(40)" 4 Cutting goal"
1040 PRINT
1050 PRINT TAB(34)"DRILL FORCES"
1060 PRINT " 5 Thrust" ;TAB(40)" 7 Horsepower"
1070 PRINT " 6 Torque"
1080 PRINT
1090 PRINT TAB(32)"SPEED AND ROTATION"
1100 PRINT " 8 Shaft speed"; TAB(40)"11 Tool feed per revolution"
1110 PRINT " 9 Surface speed"; TAB(40)"12 Tool feed per minute"
1120 PRINT "10 Rotational speed"; TAB(40)"13 Tool feed per tooth"
1130 PRINT
1140 PRINT TAB(32)"MATERIAL REMOVED"
1150 PRINT "14 Rectangular cut"; TAB(40)"16 Drill"
1160 PRINT "15 Linear feed"
1170 PRINT
1180 PRINT TAB(35)"PARTIAL CUT"
1190 PRINT "17 Distance of cut extension"
1200 PRINT
1210 INPUT "Select by number ",SELECT
1220 IF SELECT>0 AND SELECT<18 THEN RETURN
1230 PRINT "Select from 1 through 17 only"
1240 GOTO 1200
1400 '(1) Subroutine to calculate tool pressure  11111111111111
1410 GOSUB 5410               'Get constant and area
1420 P=80300!*1.33^K*A
1430 PRINT "These conditions will result in pressure"
1440 PRINT "at the tool of";P;"pounds."
1450 PRINT
1460 INPUT "Press ENTER for menu ",E$
1470 RETURN
1600 '(2) Subroutine to calculate horsepower  22222222222222
1610 GOSUB 5410               'Get constant and area
1620 INPUT "Enter cutting speed, feet per minute ",S
1630 PRINT
1640 W=2433*1.33^K*A*S
1650 PRINT "The horsepower for this cut is";W
1660 PRINT
1670 INPUT "Press ENTER for menu ",E$
1680 RETURN
1800 '(3) Subroutine to calculate machining accuracy  333333333333
1810 INPUT "Enter coefficient from Table 2-2 ",K
1820 PRINT
1830 INPUT "Enter diameter of work, inches ",D
1840 PRINT
```

Figure 2-2. (continued)

```
1850 T=K*D^.371
1860 PRINT "A reasonable tolerance for these conditions is";T;"inches."
1870 PRINT
1880 INPUT "Press ENTER for menu ",E$
1890 RETURN
2000 '(4) Subroutine to calculate cutting goal   44444444444444
2010 INPUT "Enter lower limit of acceptance range ",LL
2020 PRINT
2030 INPUT "Enter upper limit of acceptance range ",LU
2040 PRINT
2050 IF LU>LL THEN 2090
2060 PRINT "Upper limit must be larger than lower limit"
2070 PRINT
2080 GOTO 2010
2090 PRINT
2100 INPUT "Enter standard deviation ",SD
2110 PRINT
2120 PRINT "Enter cost of rejecting a piece because it is"
2130 INPUT "outside LOWER limit ",CL
2140 PRINT
2150 PRINT "Enter cost of rejecting a piece because it is"
2160 INPUT "outside UPPER limit ",CU
2170 PRINT
2180 IF CL>CU THEN 2260
2190 PRINT "The formula assumes it costs more to scrap an undersize"
2200 PRINT "piece than to rework an oversize piece.  Enter C"
2210 INPUT "to change your numbers or M to return to menu ",E$
2220 IF E$="C" THEN 2110
2230 IF E$="M" THEN 2310
2240 PRINT "Enter C or M only"
2250 GOTO 2190
2260 D=(1/(2*(LU-LL)))*(LU^2-LL^2-SD^2*LOG(CL/CU))
2270 PRINT "The machine should be set to";D
2280 PRINT
2290 INPUT "Press ENTER for menu ",E$
2300 RETURN
2400 '(5) Subroutine to calculate drill thrust  55555555555
2410 GOSUB 5610          'Get feed and diameter
2420 FTH=57.5*F^.8*D^1.8+625*D^2
2430 PRINT "Thrust for these conditions is";FTH;"pounds."
2440 PRINT
2450 INPUT "Press ENTER for menu ",E$
2460 RETURN
2600 '(6) Subroutine to calculate drill torque  666666666666
2610 GOSUB 5610               'Get feed and diameter
2620 FTQ=25.2*F^.8*D^1.8
2630 PRINT "Torque is";FTQ;"inch-pounds"
2640 PRINT
2650 INPUT "Press ENTER for menu ",E$
2660 RETURN
2800 '(7) Subroutine to calculate drill horsepower  77777777777
2810 PRINT "Type T if you have the torque; C if computer"
2820 INPUT "should calculate torque from feed and diameter ",E$
2830 PRINT
2840 IF E$="C" THEN 2880
2850 IF E$="T" THEN 2910
2860 PRINT "Respond with C or T only"
2870 GOTO 2810
2880 GOSUB 5610                'Get feed and diameter
2890 FTQ=25.2*F^.8*D^1.8
2900 GOTO 2930
2910 INPUT "Enter torque in inch-pounds ",FTQ
```

Figure 2-2. (continued)

```
2920 PRINT
2930 INPUT "Enter rotational velocity in RPM ",V
2940 PRINT
2950 FHP=15.87*10^(-6)*FTQ*V
2960 PRINT "Horsepower is";FHP
2970 PRINT
2980 PRINT "Press ENTER for menu ",E$
2990 RETURN
3100 '(8) Subroutine to calculate shaft speed in rpm  888888888888
3110 GOSUB 5710              'Get radius or diameter
3120 INPUT "Enter surface speed in feet per minute ",S
3130 PRINT
3140 V=12*S/(3.1416*D)
3150 PRINT "Shaft speed is";V;"revolutions per minute."
3160 PRINT
3170 INPUT "Press ENTER for menu ",E$
3180 RETURN
3300 '(9) Subroutine to calculate surface speed  999999999999
3310 GOSUB 5710                  'Get radius or diameter
3320 INPUT "Enter shaft rotational speed in RPM ",SR
3330 PRINT
3340 SS=.2618*D*SR
3350 PRINT "Surface speed is";SS;"feet per minute."
3360 PRINT
3370 INPUT "Press ENTER for menu ",E$
3380 RETURN
3500 '(10) Subroutine to calculate rotational speed  10 10 10 10 10 10
3510 GOSUB 5710                  'Get radius or diameter
3520 INPUT "Enter surface speed in feet per minute ",SS
3530 PRINT
3540 V=3.8197*SS/D
3550 PRINT
3560 PRINT "Rotational speed is";V;"rotations per minute."
3570 PRINT
3580 INPUT "Press ENTER for menu ",E$
3590 RETURN
3700 '(11) Subroutine to calculate tool feed per revolution  11 11 11 11 11
3710 INPUT "Enter linear feed rate, inches per minute ",AL
3720 PRINT
3730 PRINT "Type R if shaft speed is known in RPM"
3740 INPUT "or S if known in terms of surface speed ",E$
3750 PRINT
3760 IF E$="R" THEN 3800
3770 IF E$="S" THEN 3830
3780 PRINT "Respond R or S only"
3790 GOTO 3720
3800 INPUT "Enter shaft speed in rotations per minute ",SR
3810 AR=AL/SR
3820 GOTO 3860
3830 GOSUB 5710                  'Get radius of diameter
3840 INPUT "Enter shaft surface speed in feet per minute ",SS
3850 AR=.2618*D*AL/SS
3860 PRINT
3870 PRINT "Tool advance is";AR;"inches per revolution."
3880 PRINT
3890 INPUT "Press ENTER for menu ",E$
3900 RETURN
4000 '(12) Subroutine to calculate tool feed per minute  12 12 12 12 12 12
4010 INPUT "Enter tool advance rate in inches per revolution ",AR
4020 PRINT
4030 PRINT "Type R if shaft speed is known in RPM"
4040 INPUT "or S if known in terms of surface speed ",E$
```

Figure 2-2. (continued)

```
4050 PRINT
4060 IF E$="R" THEN 4100
4070 IF E$="S" THEN 4130
4080 PRINT "Respond R or S only"
4090 GOTO 4050
4100 INPUT "Enter shaft speed in rotations per minute ",SR
4110 AL=AR*SR
4120 GOTO 4160
4130 GOSUB 5710                        'Get radius or diameter
4140 INPUT "Enter shaft speed in feet per minute ",SS
4150 AL=3.8197*AR*SS/D
4160 PRINT
4170 PRINT "Tool feed is";AL;"inches per minute."
4180 PRINT
4190 INPUT "Press ENTER for menu ",E$
4200 RETURN
4300 '(13) Subroutine to calculate tool feed per tooth   13 13 13 13 13
4310 INPUT "Enter number of teeth on cutter ",T
4320 PRINT
4330 PRINT "Which set of information is known: linear feed "
4340 PRINT "rate and surface speed in rpm (type L), or feed"
4350 INPUT "in inches per revolution (type R) ",E$
4360 PRINT
4370 IF E$="L" THEN 4410
4380 IF E$="R" THEN 4460
4390 PRINT "Respond L or R only"
4400 GOTO 4320
4410 INPUT "Enter linear feed rate in inches per minute ",AL
4420 PRINT
4430 INPUT "Enter shaft speed in rotations per minute ",SR
4440 AT=AL/(T*SR)
4450 GOTO 4480
4460 INPUT "Enter tool feed rate in inches per revolution ",AR
4470 AT=AR/T
4480 PRINT
4490 PRINT "Tool feed rate is";AT;"inches per tooth."
4500 PRINT
4510 INPUT "Press ENTER for menu ",E$
4520 RETURN
4600 '(14) Subroutine -- material removed from rectangular cut   14 14 14 14
4610 INPUT "Enter width of cut, inches ",W
4620 PRINT
4630 INPUT "Enter depth of cut, inches ",H
4640 PRINT
4650 INPUT "Enter linear feed rate, inches per minute ",F
4660 PRINT
4670 VR=W*H*F
4680 PRINT
4690 PRINT "Material is removed at the rate of";VR;"cubic inches per minute."
4700 PRINT
4710 INPUT "Press ENTER for menu ",E$
4720 RETURN
4800 '(15) Subroutine -- linear feed for given material removal rate 15 15 15
4810 INPUT "Enter rate of material removal, cubic inches per minute ",VR
4820 PRINT
4830 INPUT "Enter width of cut ",W
5760 PRINT "Respond D or R only"
5770 GOTO 5710
5780 INPUT "Enter diameter in inches ",D
5790 GOTO 5820
5800 INPUT "Enter radius in inches ",R
5810 D=2*R
5820 PRINT
5830 RETURN
```

depth. When the full depth cut ends within the work, this formula also determines how far the cut extends (on the surface) beyond the point of full depth.

$$e = \sqrt{d(2r - d)}$$

where e is the extension of the cut, in inches.

Any units, inches, centimeters, or others, can be used as long as the same unit is used for all variables.

Example 2.18

A slot is to be cut $\frac{1}{2}$ inch deep and 3 inches long, using a cutter with 1-inch radius, as shown in Figure 2-1. How long will the cut be at the surface?

Length e will be at both ends of the slot, so you will double the formula and add three inches to calculate the length of the cut.

$$\text{cut} = 3 + 2 \sqrt{d(2r - d)}$$
$$= 3 + 2 \sqrt{0.5[2(1) - 0.5]}$$
$$= 4.73 \text{ inches}$$

Table 2-4 gives e for several combinations of cutter radius and depth of cut. Although entries are included for depths of cut up to the cutter's radius, a hub, mandrel, shaft, or other means of holding the cutter will usually limit the depth of cut to less than the radius.

The table's range can be extended by moving the decimal point the same number of places in every number in the table (radius, depth of cut, and body of table).

Example 2.19

A cutter with $\frac{1}{2}$-inch radius is to cut to a depth of $\frac{1}{4}$ inch. How far will it travel from first contact until it begins cutting the full $\frac{1}{4}$ inch?

To find a cutter radius of $\frac{1}{2}$ (that is, 0.5) inch, move the decimal point one place to the left in all *radius* values. Then the column headed 5 becomes the 0.5 column. It is then necessary to move the decimal point one place to the left on every number in the table, so the $\frac{1}{4}$-inch depth (that is, 0.25) is found in the first column at 2.5. At the intersection of that row and column is 4.330, which becomes 0.4330 when its decimal point is moved the same amount.

COMPUTER PROGRAM FOR SOLVING
THE FORMULAS IN THIS CHAPTER

Figure 2-2 lists a computer program that solves all the formulas in this chapter. It is structured with a subroutine for each formula, so it is easy to locate any part of the program with a typographic error. After you enter the program into your computer, test it with each of the examples.

CHAPTER 3

How to Estimate Production Time and Cost

Any production manager will agree with the old saying that "time is money." But how can you put this adage to practical use? This chapter offers a variety of formulas ranging from those that find the time for one machine to make one pass to those that project the cost and time for a project to be completed. In addition, there are calculations on line balancing, short down times, and how to allocate production time and rate. The computer program included at the end of the chapter will help you improve your scheduling procedures.

HOW TO ASSESS INDIVIDUAL MACHINE TIME REQUIREMENTS

These formulas give information about time requirements for individual production machines.

Symbols used

c_p = depth of cut made by one pass, inches

c_t = total depth of cut, inches

d_f = diameter of stock at finish, inches

d_s = diameter of stock at start, inches

f_r = feed, in inches per revolution of spindle

l = length of cut, inches

l_e = entry and leaving length, inches

l_w = length of work, inches

m = movement of base of milling machine, inches

r = rotational speed, revolutions per minute

t = time, minutes

t_c = time to make cut, minutes

t_p = length of time for one pass, minutes

t_t = total time for machining, minutes

Figuring Time per Pass on Automatic Feed Machines

When the tool travels on automatic feed, the following formula gives the time required for one pass over the work.

$$t_p = \frac{l_w + l_e}{f_r r}$$

Example 3.1

A solid metal cylinder with $1\frac{1}{4}$-inch diameter is to be turned down to a smaller diameter for $3\frac{1}{2}$ inches of its length. The tool will start $\frac{1}{8}$ inch from the work, end on the work, and feed .018 inch per revolution. How long will one pass take if the lathe is turning 733 rotations per minute?

$$t_p = \frac{3.5 + 0.125}{0.018(733)}$$

$$= 0.27 \text{ minute for one pass}$$

Determining Time per Cut

This formula finds total machining time by multiplying time per pass by the number of passes. It does not include time for the tool to return for another pass.

$$t_c = \left(\frac{l_w + l_e}{f_r r}\right)\left(\frac{c_t}{c_p}\right)$$

Example 3.2

In the preceding example each pass cuts 0.003 inch. How long will it take to cut to a depth of 0.130 inch?

$$t_c = \left(\frac{3.5 + 0.125}{0.018(733)}\right)\left(\frac{0.130}{0.003}\right)$$

$$= 11.7 \text{ minutes}$$

Calculating Length of Time per Piece

When cylindrical stock is turned down from one diameter to another, this formula finds the length of time for the total operation, exluding time required to return the tool after each pass, unless that time is included in the time for a pass.

$$t_t = \frac{t_p}{2c_p}(d_s - d_f)$$

Example 3.3

A cylinder is to be turned from $1\frac{1}{4}$ inches to $\frac{7}{8}$ inch diameter, using cuts of 0.012 inch. How long will it take if each pass takes 0.33 minutes?

$$t_t = \frac{0.33}{2(0.012)}(1\tfrac{1}{4} - \tfrac{7}{8})$$

$$= 5.16 \text{ minutes}$$

Time Requirements for Completing a Milling Operation

This formula gives the time to complete a milling operation, given the length of cut and the table feed rate.

$$t = \frac{l}{m}$$

Because of its simplicity, the formula is easily adapted to a variety of units. Table 3-1 lists some examples of combinations for which the length dimensions are the same.

Example 3.4

How long will it take a machine to mill 20 centimeters if the feed rate is 0.05 centimeters per second?

$$t = \frac{20}{0.05}$$

$$= 400 \text{ seconds}$$

Metric versus U.S. Measurements: Formulas for Coping With Mixed Units

As the United States is in transition to the metric system, it is common to see machines using one set of dimensions and the work using another. If length of cut is in centimeters and table feed rate is in inches, the formula becomes

$$t = 0.3937 \left(\frac{l}{m}\right)$$

and if length of cut is in inches and table feed rate is in centimeters, the formula becomes

$$t = 2.54 \left(\frac{l}{m}\right)$$

Table 3-2 gives combinations of mixed units that can be applied to these formulas.

Table 3-1.   Combinations of units having the same length dimension.

l	*m*	*t*
in.	in./min.	min.
in.	in./sec.	sec.
cm.	cm./min.	min.
cm.	cm./sec.	sec.

Table 3-2. Combinations of units with different length dimensions.

l	m	t
cm.	in./min.	min.
cm.	in./sec.	sec.
in.	cm./min.	min.
in.	cm./sec.	sec.

Example 3.5

How long will it take to mill 13 centimeters if the table feeds at 0.018 inch per second?

$$t = 0.3937 \left(\frac{13}{0.018} \right)$$

$$= 284 \text{ seconds}$$

FORMULAS FOR DEALING WITH OVERALL PRODUCTION MAINTENANCE

This section moves from individual machines to some formulas that apply to the overall production process.

Symbols used

c = cost to produce one unit

c_1 = cost of material for one unit

c_i = indirect costs allocated to one unit

c_m = machine operating costs allocated to one unit

c_p = periodic costs allocated to one unit

r = production rate, pieces per unit of time

t = total machine time allocated to one unit, minutes

t_1 = machine time to produce one unit, minutes

t_a = periodic down time allocated to one unit, minutes

t_r = time to replace worn tool, minutes

t_s = set-up time for one unit, minutes

t_t = time between periodic short down times, minutes

v_s = value of scrap

Line Balancing: How to Find the "Right" Time for Sequential Operations

When the production process consists of sequential operations by different people or stations, the ideal situation is for each operation to take exactly the same time. Operations that require less time than others result in slack time.

The Equalization Factor

Equalization factor e is a measure of balance; it can have values from zero to one, with one being perfect balance.

$$e = 0.0167r \sum_{i=1}^{n} t_i$$

where $\sum_{i=1}^{n} t_i$ means the summation of values with subscripts 1 through n.

Example 3.6

An assembly line consisting of 8 stations completes 14 items an hour. A thorough time study of the operations reveals that they should take 28, 30, 28, 32, 31, 29, 28, and 29 seconds.

As t_i is defined in minutes, we divide each of the station times by sixty and then substitute directly into the formula.

$$e = 0.0167(14) \sum_{i=1}^{8} t_i$$

$$= 0.0167(14) \left(\frac{28}{60} + \frac{30}{60} + \frac{28}{60} + \frac{32}{60} + \frac{31}{60} + \frac{29}{60} + \frac{28}{60} + \frac{29}{60} \right)$$

$$= 0.9157$$

The individual operations add up to 3.9167 minutes but the production rate is one every 4.2857 minutes. As these two numbers approach each other, the equilization factor approaches one.

Short Down Times: Calculating Necessary Periods of Inoperation

Considered here are short periods of inoperation, as when the operator changes a tool, checks settings, or lubricates a point. For thorough cost accounting, these times should be allocated to pieces produced according to the following formula.

$$t_a = \frac{t_r t_1}{t_t}$$

Keeping in mind the purpose of this calculation — to allocate short periods of inoperation — helps determine what to include in t_a. For example, consider a worn tool; it must be removed, sharpened, and replaced. If, during production, the operator periodically removes a worn tool and inserts another from a stock of sharpened ones, that time is part of t_a. If the sharpening is done at another time, instead of while the machine is stopped between production units, that time is not included.

Example 3.7

In a critical phase of production the operator stops after 12 minutes and checks the limit settings on the machines. It takes 24 seconds to make the check. How much of this down time should be allocated to each piece produced if each piece is machined for 53 seconds?

As most symbols are defined in minutes, divide the figures given in seconds by sixty.

$$t_a = \frac{\left(\dfrac{24}{60}\right)\left(\dfrac{53}{60}\right)}{12}$$

$$= 0.0294 \text{ minute}$$

When it is more convenient to work in seconds, then, instead of converting the 24 and 53 seconds into minutes, we convert the 12 minutes into seconds by multiplying by 60.

$$t_a = \frac{24(53)}{12(60)}$$

$$= 1.77 \text{ seconds}$$

Allocating Production Time and Rate

Total operating machine time to be allocated to one piece is the sum of allocations for short down time, individual setup time, and actual machining time.

$$t = t_a + t_s + t_1$$

As t is in minutes per piece, its reciprocal will find pieces per minute, the production rate.

$$r = \frac{1}{t}$$

Example 3.8

In the preceding example it takes 18 seconds to remove a finished piece, place new work, and bring the tool to position. How much machine time should be allocated to each piece, and what is the production rate?

$$t = 0.0294 + \frac{18}{60} + \frac{53}{60}$$

$$= 1.2127 \text{ minutes per piece}$$

or

$$t = 1.77 + 18 + 53$$

$$= 72.77 \text{ seconds per piece}$$

The production rate is

$$r = \frac{1}{1.2127}$$

$$= 0.8246 \text{ piece per minute}$$

How to Measure Unit Cost Based on Four Main Components

This formula combines several costs associated with the production of one unit. First the formula is given, then the meaning of several of its components is discussed.

$$c = c_m + c_p + c_i + c_1 - v_s$$

Cost of operating machine

This item is the sum of costs that can be directly identified to the production of one piece. Most prominent is the labor for time t. If depreciation is recorded on a per-piece basis, it can be included.

Periodic costs

If a cutting tool is replaced after a given number of pieces, the amount for each piece is included here. For tools that are sharpened or otherwise restored, this cost includes allocation of labor and machines for restoring.

Indirect costs

A certain portion of the production manager's salary, as well as other management and support personnel salaries can be assigned to each piece. In addition, facilities costs, depreciation, heat, rent, and many other overhead costs are included as indirect costs. In some industries, these costs can exceed direct labor costs.

Cost of material

This item is the net of material used, material scrapped, shrinkage, and minus salvage value of material not used. Hauling, storing, handling, and other charges are assumed to be included if they can be identified. Purchase and inventory costs are also included. As the formula calculates cost, the revenue received from selling scrap is entered as a negative number.

Example 3.9

When purchasing, shipping, and other costs are included, a firm pays $1,325.00 for the material from which it produces 500 units. After 3,000 units, they sell the shavings for $11.00. The production machine uses a throwaway cutter that costs $2.37 and is replaced every 40 pieces. Indirect charges of $109.00 are attached to every 1,000 pieces. The direct labor rate is $31.22 per hour, and 2.5 minutes of operation time are required for each item. What is the cost per piece for this operation?

$$c = 31.22 \left(\frac{1}{60}\right) 2.5 + 2.37 \left(\frac{1}{40}\right) + 109 \left(\frac{1}{1,000}\right) + 1,325 \left(\frac{1}{500}\right) - 11\left(\frac{1}{3,000}\right)$$

$$= \$4.12$$

SCHEDULING: HOW TO FIGURE ESTIMATED COMPLETION TIME BASED ON PERFORMANCE TO DATE

This section gives some formulas that make projections based on performance to date. The formulas project costs and time, and give statistical probability of a given completion time. A computer program is included to solve all the formulas.

Symbols used

E_c = expected time of critical activity

E_t = expected time to complete

e_b = best, or most optimistic, estimate

e_l = estimate of most likely time

e_w = worst case, or most pessimistic, estimate

t = an estimated time to complete

t_m = mean time estimate of time to complete

v = variance

z = number of standard deviations

σ = standard deviation

Tips on Providing an Unbiased Best Estimate of Completing an Activity

A *best estimate* of time to complete a project implies *most likely*, and you might expect that best estimates would err on the short side as often as on the long side. However, the human tendency is to cover for contingencies, producing a "best estimate" that is biased on the long side. A successful method of solving this problem is to have three estimates prepared — worst case (e_w), best case (e_b), and most likely (e_l). It is then easier for the estimator to provide an unbiased best estimate.

The worst case should not be an everything-goes-wrong estimate; it is sometimes called a ninety-nine percent estimate. That is, there is a ninety-nine percent probability the actual time will not exceed this estimate. It should be thoughtfully made with this rule in mind.

The best case estimate is a one percent estimate; there is less than one percent probability the actual time will be less. Now the planner can make an unbiased most likely estimate.

It is not necessary that the most likely estimate be centered between the other two, or that there be any symmetry at all. Therefore a fourth value is brought in — the *expected value*, which will differ from the most likely if the three estimates are not symmetrical. It is found with a formula developed by statisticians.

$$E_t = \frac{e_w + 4e_l + e_b}{6}$$

We define *expected value* as the average of outcomes if we applied these rules a very large number of times.

Example 3.10

An experienced estimator makes one- and ninety-nine-percent estimates of 5.6 and 13.4 hours respectively for completing a certain activity. The most likely estimate is 7.7 hours. What is the expected time to completion?

$$E_t = \frac{13.4 + 4(7.7) + 5.6}{6}$$

$$= 8.3 \text{ hours}$$

Estimating Mean Time Completion for a Project

To determine completion time for a project, first determine which activities are critical. That is, which activities will affect the overall schedule if delayed. For example, Figure 3-1 shows two parallel paths of activities that must be completed before the nameplate can be affixed. If the top path takes three hours and the bottom one two days, then a delay in the bottom one will delay application of the nameplate and everything after it. A delay in assigning the serial number and stamping the nameplate will not affect the overall schedule. Therefore painting and drying is a critical activity.

Estimated Mean Time

Estimated mean time for a project is the sum of the expected times of critical activities.

$$t_m = \sum E_c$$

Example 3.11

The critical activities of a short project have expected completion times of 6, 5, 8, 4, and 6 days. What is the mean time estimate of this project?

$$t_m = 6 + 5 + 8 + 4 + 6$$

$$= 29 \text{ days}$$

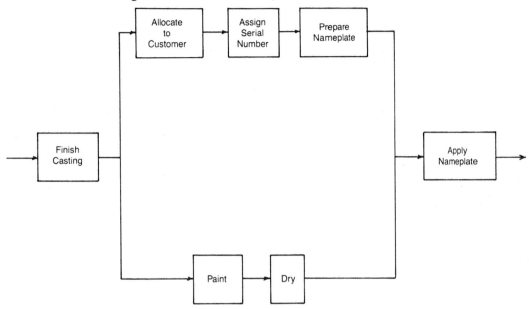

Figure 3-1. The critical path is the longer one.

Variance Estimating: Determining the Band of Possibilities

Variance is a measure of how widely scattered the possibilities can be. (See Chapter 11 for more information on this subject.)

$$v = \left(\frac{e_w - e_b}{6} \right)^2$$

Example 3.12

The worst case estimate for lacing an electrical harness is 4.2 hours and the best case estimate is 3.7 hours. What is the variance of this activity?

$$v = \left(\frac{4.2 - 3.7}{6} \right)^2$$

$$= 0.0069 \text{ hours}$$

Standard Deviation: Figuring the Square Root of Variance

Standard deviation, the square root of variance, is a more popular measure of dispersion. (See Chapter 12.) For an entire project, standard deviation is the square root of the sum of variances for the critical activities.

$$\sigma = \sqrt{v_{\text{critical}}}$$

Example 3.13

Critical activities of a project have variances of 0.0069, 0.0130, 0.0092, and 0.0114 hours. What is the standard deviation of this short project?

$$\sigma = \sqrt{0.0069 + 0.0130 + 0.0092 + 0.0114}$$

$$= 0.2012 \text{ hours}$$

Probability of Other Completion Times

The preceding items plus the normal area table (Table A-1 in the appendix) are used to find the probability of the project's being completed within a given length of time. First find how many standard deviations (z) the time is from mean.

$$z = \frac{t - t_m}{\sigma}$$

Then use the normal area table to find the probability of that number of standard deviations.

Example 3.14

The mean estimate for a project completion is 205 days, with a standard deviation of 14. A bonus will be paid if completion is less than 190 days, and a penalty will be assessed if completion is later than 215 days. Knowing the probabilities of each of these completions is important to those negotiating the price of this contract.

$$z_{early} = \frac{190 - 205}{14}$$

$$= -1.07$$

$$z_{late} = \frac{215 - 205}{14}$$

$$= 0.71$$

Look at the positive (late) figure first. The negotiators want to know the probability of the completion time being more than 0.71 standard deviations from mean. Table A-1 gives probabilities of being to the right of mean; the probability of between zero and 0.71 standard deviations to the right is 0.2611. Therefore, the probability of being to the *right of 0.71* is 0.5 − 0.2611, or 0.2389. There is a 23.89 percent probability the project completion will be later than 215 days.

For the probability of finishing early, picture Table A-1 reversed. The negotiators want to know the probability of being *more than* 1.07 standard deviations to the left of mean. From the table, there is a 0.3577 probability of between zero

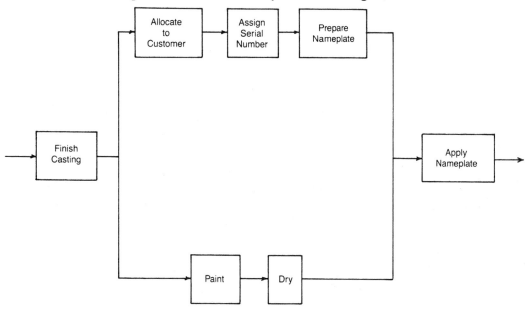

Figure 3-2. Computer program for schedule calculations.

```
10 '****************************************************************
20 '** SCHED.BAS  Does scheduling formulas from Chapter 3 of   **
30 '** Handbook of Manufacturing and Production Management     **
40 '** Formulas, Charts, and Tables                            **
50 '****************************************************************
60 CLEAR,,,32768!: SCREEN 6: COLOR 3,2
70 DIM A(100,6)
80 UI$="###": UD$="#####.####": U2$="#.##"
90 CLS
100 GOSUB 160                      'Display menu
110 CLS
120 ON SELECT GOSUB 310,610,790,870,1000
130 GOTO 90
140 END
150 'Subroutine to display menu and accept selection  <<<<<<<<<<<<<
160 PRINT
170 PRINT TAB(25) "DATA INPUT/OUTPUT"
180 PRINT "1 Enter estimates from keyboard
190 PRINT "2 Retrieve estimates from disk"
200 PRINT "3 Save estimates now in memory"
210 PRINT TAB(23) "USE OF DATA IN MEMORY"
220 PRINT "4 Display table"
230 PRINT "5 Calculate probabilities"
240 PRINT
250 INPUT "Select by number ",SELECT
260 IF SELECT>0 AND SELECT<6 THEN RETURN
270 PRINT
280 PRINT "Select only a number from 1 through 6"
290 GOTO 240
300 '(1) Subroutine to enter data from keyboard  111111111111
310 N=0: SUMSIG=0: TMEAN=0
320 PRINT
330 PRINT TAB(20) "ACTIVITY #";N+1
340 INPUT "Enter 'worst case' estimate ... or END ",E$
350 IF E$="END" THEN 580
360 N=N+1
370 A(N,1)=VAL(E$)
380 INPUT "Enter 'most likely' estimate ",A(N,2)
390 IF A(N,2)=<A(N,1) THEN 430
400 PRINT "Most likely cannot be longer than worst case!!!"
410 N=N-1
420 GOTO 320
430 INPUT "Enter 'best case' estimate ",A(N,3)
440 IF A(N,3)<=A(N,2) THEN 480
450 PRINT "Best case cannot be longer than most likely!!!"
460 N=N-1
470 GOTO 320
480 A(N,4)=(A(N,1)+4*A(N,2)+A(N,3))/6          'Expected value
490 A(N,5)=(A(N,1)-A(N,3))^2/36                'Variance
500 INPUT "Is this a critical activity?  (Y/N) ",Q$
510 IF Q$="Y" THEN A(N,6)=1: GOTO 550
520 IF Q$="N" THEN A(N,6)=0: GOTO 320
530 PRINT "Respond Y or N only"
540 GOTO 500
550 TMEAN=TMEAN+A(N,4)
560 SUMSIG=SUMSIG+A(N,5)
570 GOTO 320
580 SIGMA=SQR(SUMSIG)
590 RETURN
600 '(2) Subroutine to retrieve data from disk  2222222222222
610 INPUT "What is name of file on disk? ",N$
620 OPEN "I",#1,N$
630 N=0: SUMSIG=0: TMEAN=0
640 IF EOF(1) THEN 680
650 N=N+1
```

Figure 3-2. (continued)

```
660 INPUT #1,A(N,1),A(N,2),A(N,3),A(N,6)
670 GOTO 640
680 CLOSE
690 FOR C=1 TO N
700    A(C,4)=(A(C,1)+4*A(C,2)+A(C,3))/6          'Expected value
710    A(C,5)=(A(C,1)-A(C,3))^2/36                'Variance
720    IF A(C,6)=0 THEN 750
730    TMEAN=TMEAN+A(C,4)
740    SUMSIG=SUMSIG+A(C,5)
750 NEXT C
760 SIGMA=SQR(SUMSIG)
770 RETURN
780 '(3) Subroutine to save data on disk   3333333333333
790 INPUT "What name will be used to save data? ",N$
800 OPEN "O",#2,N$
810 FOR C=1 TO N
820    WRITE #2,A(C,1),A(C,2),A(C,3),A(C,6)
830 NEXT C
840 CLOSE
850 RETURN
860 '(4) Subroutine to display table   4444444444444444
870 PRINT "Activity";TAB(13)"Worst";TAB(25)"Likely";TAB(38)"Best";TAB(48)"Expect
ed";TAB(60)"Variance";TAB(72)"Critical"
880 FOR C=1 TO N
890    PRINT TAB(3) USING UI$;C;
900    FOR D=1 TO 5
910       PRINT TAB(12*D-1) USING UD$; A(C,D);
920       IF A(C,6)=0 THEN CR$=" no" ELSE CR$="yes"
930    NEXT D
940    PRINT TAB(74) CR$
950 NEXT C
960 PRINT "Project standard deviation =";SIGMA
970 INPUT "Press ENTER to return to menu ",E$
980 RETURN
990 '(5) Subroutine to calculate probabilities   5555555555555
1000 PRINT "Type E if you will enter data here, or"
1010 INPUT "M if data in memory is to be used ",E$
1020 PRINT
1030 IF E$="E" THEN 1070
1040 IF E$="M" THEN 1100
1050 PRINT "Type E or M only"
1060 GOTO 1000
1070 INPUT "Enter project mean estimate ",PME
1080 INPUT "Enter project standard deviation ",PSD
1090 GOTO 1120
1100 PME=TMEAN
1110 PSD=SIGMA
1120 INPUT "You want probability of completion in how long? ",T
1130 Z=(T-PME)/PSD
1140 PRINT "Enter the value for ";
1150 PRINT USING U2$;ABS(Z);
1160 PRINT " from Table A-1";
1170 INPUT " ",TABLE
1180 IF Z<0 THEN CD$=" earlier " ELSE CD$=" later "
1190 PRINT "Probability of completion";CD$;"than";T;"is";.5-TABLE
1200 PRINT
1210 PRINT "Type D for probability of another data, keeping the"
1220 PRINT "mean and standard deviation; N for a probability with"
1230 PRINT "new data; or M to return to menu ";
1240 INPUT " ",E$
1250 IF E$="D" THEN 1120
1260 IF E$="N" THEN 1000
1270 IF E$="M" THEN RETURN
1280 PRINT "Respond with D, N, or M only"
1290 GOTO 1200
```

Figure 3-2. (continued)

```
4840 PRINT
4850 INPUT "Enter depth of cut ",H
4860 PRINT
4870 F=VR/(W*H)
4880 PRINT "Feed rate is";F;"inches per minute"
4890 PRINT
4900 INPUT "Press ENTER for menu ",E$
4910 RETURN
5000 '(16) Subroutine to calculate material removed by drill  16 16 16 16
5010 PRINT "Enter angle of drill tip.  Use 180 if tip correction"
5020 INPUT "is not to be used ",T
5030 PRINT
5040 INPUT "Enter depth of hole ",H
5050 PRINT
5060 INPUT "Enter diameter of drill ",D
5070 PRINT
5080 V=(3.1416*D^2/4)*(H-(D/3)*(1/TAN((T/2)*3.1416/180)))
5090 PRINT "Volume of material removed by drill is";V;"cubic inches."
5100 PRINT
5110 INPUT "Press ENTER for menu ",E$
5120 RETURN
5200 '(17) Subroutine to calculate cut extension   17 17 17 17 17 17
5210 INPUT "Enter depth of cut ",D
5220 PRINT
5230 INPUT "Enter radius of cutter ",R
5240 PRINT
5250 E=SQR(D*(2*R-D))
5260 PRINT "Cut will extend";E;"inches on surface."
5270 PRINT
5280 INPUT "Press ENTER for menu ",E$
5290 RETURN
5400 'Subroutine to enter constant and area   <<<<<<<<<<<<<<<
5410 PRINT "Type A if area of chip is known, or D if you"
5420 INPUT "have depth of cut and feed rate ",E$
5430 PRINT
5440 IF E$="A" THEN 5490
5450 IF E$="D" THEN 5510
5460 PRINT "Respond A or D only"
5470 PRINT
5480 GOTO 5410
5490 INPUT "Enter area of chip in square inches ",A
5500 GOTO 5550
5510 INPUT "Enter depth of cut in inches ",D
5520 PRINT
5530 INPUT "Enter feed rate in inches per revolution ",F
5540 A=D*F
5550 PRINT
5560 RETURN
5600 'Subroutine to enter feed and diameter   <<<<<<<<<<<<
5610 INPUT "Enter feed rate, inches per rotation ",F
5620 PRINT
5630 INPUT "Enter tool diameter, inches ",D
5640 PRINT
5650 RETURN
5700 'Subroutine to enter radius or diameter   <<<<<<<<<<<<
5710 PRINT "Type D if you are going to give tool diameter,"
5720 INPUT "or R if you are going to give radius ",E$
5730 PRINT
5740 IF E$="D" THEN 5780
5750 IF E$="R" THEN 5800
```

Table 3-3. Schedule for a short project.

Activity	Worst	Likely	Best	Expected	Variance	Critical
1	14.0000	12.0000	10.0000	12.0000	0.4444	yes
2	12.0000	8.0000	6.0000	8.3333	1.0000	yes
3	9.0000	7.0000	6.0000	7.1667	0.2500	yes
4	12.0000	10.0000	8.0000	10.0000	0.4444	no
5	13.0000	11.0000	8.0000	10.8333	0.6944	yes
6	8.0000	6.0000	4.0000	6.0000	.04444	yes

Project standard deviation = 1.683251
Press ENTER to return to menu

and 1.07 standard deviations, so there is $0.5 - 0.3577$ or 0.1423 probability of being to the left of 1.07. There is a 14.23 percent probability of earning the bonus for being early.

COMPUTER PROGRAM

Scheduling can involve many activities, and the calculations shown here can be a burden. The computer program listed in Figure 3-2 performs the calculations and prints a table showing all values. It also finds probabilities of different completion dates.

Example 3.15

Table 3-3 is a print of the screen for a short project with six activities, five of them critical.

After you key the program into your computer, select number one from the menu and then enter the numbers shown in the worst, likely, and best columns of Table 3-3. When the screen asks for activity number seven, type **END** and the menu will return. Now select four, and the screen shown in Table 3-3 will appear.

After examining the table, press **ENTER** to return to the menu. Then select five and enter the numbers given in the probability example just before the computer program. The computer should show the same results as in the example.

How to Measure Worker Productivity With Controlled Observation Samplings

The entire field of worker measurement is based on sampling. That is, projections or conclusions are based on observations of less than 100 percent of the activity being measured. But whether the sample is taken in the production department or elsewhere, it is always necessary to know how many observations are required for a given level confidence in the conclusions. This chapter looks at some of the different types of observation samplings possible, including binary observation, random work samplings, and the traditional stopwatch method. In addition, you'll learn how to determine a performance rating based on the workers' reactions to being observed and how to measure "normal" and standard time.

BINARY OBSERVATIONS: A RESTRICTIVE SAMPLING FOR MEASURING TIME

In this most restrictive sampling method, the observer looks for one of two possibilities. The purpose is to determine the percentage of time that a certain condition exists. For example, the study could determine how long a worker spends at a certain task, or how long a worker waits for a turn on a time-share terminal. Table 4-1 lists some pairs of activities that could be observed.

Binary observations recognize only two possibilities. For example, in number 1 of Table 4-1, if the worker is not at the bench the observer does not record exactly where the worker is.

Number 4 is poorly written because it does not include every possibility in the two choices. What should the observer record if the worker is kneeling? All the other examples in Table 4-1 follow the safe method of naming an activity and then letting the other choice be simply "not" that activity.

Drawing Conclusions from Observations

After a number of observations have been made, you can tentatively conclude that the percentage of times observed at the activity is the same as the percentage of time actually spent at the activity,

$$p = \frac{n_0}{n}$$

where

p = portion of observations at activity

n_0 = number of observations in which worker was doing indicated activity

n = total number of observations

Multiply ratio p by 100 to express it as a percentage.

Example 4.1

A drill press operator is being observed for "drill in contact with work" and "not in contact." In 125 observations the drill was in contact 25 times.

Table 4-1. Typical two-state observations.

1. Worker is at bench	Not at bench
2. Worker has machine running	Not running
3. There is a backlog by the worker	No backlog
4. Worker is sitting	Standing
5. Worker is using pliers	Not using pliers
6. Worker is using computer tape	Not using tape

$$p = \frac{25}{125}$$

$$= 0.2 \text{ or } 20 \text{ percent}$$

The observer concludes that, as the drill was in contact with the work 20 percent of the times the operator was checked, it was in contact 20 percent of the time the operator was working.

Closeness Factor: A Compromise Between Observed and Actual Ratios

It is always possible that the observer, by chance, found the drill in contact with the work more or less than a proportionate number of times. As the checks were random, it is possible—even likely—that, although the drill was in contact with the work twenty percent of the times the observer checked, the actual percent of time in contact was different. A closeness factor, c, is defined as the percentage or fractional difference between observed ratio and actual ratio.

Example 4.2

Fifteen percent of an inspector's time is spent writing down test results. Then a closeness factor of five percent means that fifteen percent plus or minus five percent of the observations should find the inspector writing down test results. Five percent of fifteen is

$$.05(15\%) = 0.75\%$$

and the limits are

$$15\% - 0.75\% = 14.25\%$$

$$15\% + 0.75\% = 15.75\%$$

If the true ratio is 15 percent, then an observed ratio between 14.25 and 15.75 percent is within five percent of actual.

We will interpret the closeness factor loosely. That is, if the observed ratio is p, we will say that the actual ratio is within c percent of p. We will also say it could mean that, if the actual ratio is r, the observed ratio should be within c percent of r. Of course, we will more often use an observed ratio to estimate the actual ratio. However, as we shall see, a preliminary estimate of the actual ratio is necessary in order to determine how many observations should be made.

Closeness factor is a compromise—we would like to know within narrow bounds what the actual ratio is, but the number of observations required increases as the bounds are narrowed.

Degree of Confidence: The More You Observe, the More Likely You Are to Be Accurate

Every worker has probably said, "I stop work for one second and that's when the boss walks in." An experienced boss knows, too, that one observation does not give a reliable indication of what the worker does.

But if the boss walks in ten times and finds the employee not working, he or she has more reason to question the employee's effort. One hundred such observations increase the confidence in that conclusion.

The confidence level is therefore related to the number of observations. It answers the question, "How sure are we that the observed ratio is within c percent of the actual ratio?" If, for example, you chose a 95 percent confidence level, you mean that 95 times out of 100 the observed ratio is within c percent of the actual ratio.

Degree of confidence, like closeness ratio, is a compromise: high levels of confidence require larger numbers of observations. Confidence factor, z, is found as shown in the following example.

Example 4.3

Find the confidence factor for a 96 percent level of confidence.

1. Divide the confidence level by two. If it is given as a percentage, first convert it to a decimal. Ninety-six percent becomes 0.96, then divide it by two to get 0.4800.

2. Turn to Table A-1 in the appendix and look through the body of the table for the nearest entry to the number just found. The nearest to 0.4800 is 0.4798.

3. Read the headings of the row and column that intersect there; z is their sum. The row that contains 0.4798 is headed by 2.0 and the column is headed by .05. Therefore z for 96 percent is 2.05.

Required Number of Observations: Determining the Minimal Amount Necessary

Closeness and confidence factors can be combined in a formula that shows the minimum number of observations that should be made.

$$n = \frac{z^2 (1 - r)}{c^2 r}$$

where

n = minimum number of observations

z = confidence factor

r = estimate of the actual ratio

c = closeness factor

As n depends on the actual value of r, but r is the figure you are trying to determine, the usual procedure is as follows:

1. Make a preliminary estimate of r.
2. Use that estimate to calculate n.
3. Make about sixty to seventy-five percent of the indicated number of observations.
4. Calculate r.
5. Use that r to calculate a new n.
6. If the new n is higher than the number of observations already made, continue making observations.

Example 4.4

The production department is preparing an upgrade kit for an electronic amplifier they sold until last year. One item in the kit is a forty-seven-inch length of hookup wire. Each worker packing the kits has been told to estimate the length, but to measure the estimated length every tenth kit. This plan is to guard against a slow drift in what packers perceive as forty-seven inches.

After a while an observer is sent to see how the packers are following instructions. Because exact compliance is not critical, the observer will be satisfied with being eighty percent confident that packers are checking their estimated lengths within thirty percent of every tenth. Thirty percent of ten is three, so the observer will be satisfied if packers are checking every seventh, eighth, ninth, tenth, eleventh, twelth, or thirteenth kit. For each packer, the observer notes either, "checked length on this kit" or "did not check length on this kit."

Half of 0.80 (80 percent) is 0.40, and the nearest to it in Table A-1 is 0.3997, which is at the intersection of 1.2 (row) and .08 (column). The confidence factor is their sum, or $z = 1.28$. Substituting that value, along with $r = 0.10$ (a worker following instructions exactly will check 1/10 of the kits packed) and $c = 0.30$, gives

$$n = \frac{1.28^2 (1 - 0.1)}{0.30^2 (0.1)}$$

$$= 164$$

If the sampling ratio equals the assumed ratio we used for this calculation (0.10), then 164 observations will verify this fact with a confidence level of eighty percent. However, if the observed ratio is different from the assumed ratio, then a new n should be calculated. Most experienced observers will stop for a tentative calculation of r and a new calculation of n before making all the observations.

For example, let us say that, after 120 observations, one packer has checked the measurement 11 times. The new tentative r is now 11/120, or .0917. As it is different from the r used to calculate n, the observer calculates a new n.

$$n = \frac{1.28^2 (1 - 0.0917)}{0.30^2 (0.0917)}$$

$$= 180$$

The observer makes 60 more observations, for a total of 180. If the packer measures 6 times in those 60 observations, the final r will be 17/180 or .0944. As r for one out of 10 is 0.1000 and for 1 out of 11 is .0909, this packer averages a measurement after every 10 or 11 kits.

How to Proceed When the Confidence Level Is Unknown

Instead of deciding on the desired confidence level and determining how many observations to make, you might take a certain number of observations and then want to know the confidence level they give you. This situation could arise if observations were taken during a preproduction run that made samples for buyers. During the two weeks before production is resumed, you want to adjust the assembly line's balance; no changes are to be made after that.

The formula for confidence factor is

$$z = c \sqrt{\frac{nr}{1 - r}}$$

Example 4.5

A runner loads raw material into a cart and delivers it to work stations. In 210 observations, an observer notes the runner "loading the cart" 30 times and "not loading the cart" 180 times. If the observer could be reasonably confident that the runner spends between ten and twenty percent of his time loading the cart, she will terminate the observations, write her report, and move on to another assignment.

Ten to twenty percent is the same as fifteen percent plus or minus five percent. Five percent is one-third or 33.33 percent of fifteen percent. Therefore the observer will substitute 0.3333 for c. For r she will substitute 30/210, or 0.1429.

$$z = 0.3333 \sqrt{\frac{210(0.1429)}{1 - 0.1429}}$$

$$= 1.9722$$

Table A-1 provides for only two decimal places, so the observer uses 1.97. She goes down the left-hand column to 1.9, across to the column headed .07, and reads .4756. This is the reverse of the steps in the last example, so the observer *doubles* 0.4756. The observer is now 95.12 percent confident that the runner spends between ten and twenty percent of his time loading the cart.

WORK SAMPLING: RANDOM OBSERVATIONS WITHOUT RESTRICTIONS

Here the observer has a prepared list, with no limit on the number of items that can be included. At random times the observer notes simply *which* of the activities the worker is doing. It is assumed that if, for example, sixteen percent of the observations show the worker testing diodes, then close to sixteen percent of that worker's day is spent testing diodes.

One advantage to obtaining data through work sampling is that it does not require a highly experienced observer. It also avoids the distortions found when the constant presence of an observer affects a worker's routine (see PERFORMANCE RATING in this chapter). Another advantage is that one observer can sample the work of many individuals or groups as fast as one worker could be timed.

If just the fraction of a day at a certain task is wanted, instead of length of time, the formula is

$$r = \frac{n_0}{n_t}$$

This ratio is part of the formula for time on a task

$$t_0 = rt_s$$

where

t_0 = time worker spends on task 0, minutes

t_s = length of shift, minutes

n_0 = number of times worker was observed at task 0

n_t = total number of times worker was observed

Example 4.6

A lathe operator's activities are noted 240 times during an eight-hour shift. Sixteen of the 240 observations found the worker cutting off a section with a parting tool. How many minutes a day does the worker probably spend on that part of the job?

As t_s must be in minutes in the formula, you include the factor sixty to convert eight hours to minutes.

$$t_0 = 8(60) \left(\frac{16}{240} \right)$$

$$= 32 \text{ minutes}$$

Closeness and Confidence Factors to Consider

As with binary observations, you must consider closeness factor, confidence factor, and the number of observations necessary.

$$n = \frac{z^2 (1 - r)}{c^2 r}$$

where all symbols have the same meaning they had for binary observations. Factors z and c are determined in the same way.

Although binary observations are a restricted form of work sampling, we explain them separately because there are more issues to consider when several tasks are being examined. Each task probably takes a different length of time and therefore requires a different minimum number of observations. It is also possible that some tasks are more critical than others, so their times should be known to a higher degree of accuracy and with more confidence.

In the extreme there could be a different r, c, and z for each task. Then n would have to be calculated for each task, and the largest would be the one used.

But, realistically, an analyst scans the list of tasks, recognizing that the smallest r, smallest c, and largest z require the largest n. The analyst can then calculate n for one or two critical tasks.

COMPUTER PROGRAM FOR DETERMINING NECESSARY AMOUNT OF OBSERVATIONS

Figure 4-1 lists a computer program that does all the work of calculating n for each task. It then displays a table so that the observer can determine how many observations to make.

The program gives the observer a choice when entering data. He or she can enter a different closeness factor for each activity, or use the same factor for all. The same choice can be made for the confidence factor. When the same factor for all is chosen, the computer will ask for its value before asking for data—otherwise the factors will be requested along with data.

Example 4.7

Use the program to plan the observations listed in Table 4-2. Although this list contains an unlikely mixture of closeness and confidence factors, it demonstrates the ability of the computer program to calculate n for each task.

Key the program into your computer. After you proofread the program, run it. For computers that do not have 80-column screens, it may be necessary to adjust print positions in lines 2410, 2420, 2430, and 2480 through 2540.

The program avoids scrolling by printing twenty lines at a time, then waiting until the operator is ready to view the next twenty lines. If some other number of lines is better for your computer, adjust lines 2450, 2570, and 2580, **CLS** is a command to clear the screen; **HOME**, **CALL CLEAR**, or other commands should be used for some computers. Except for those items, the commands and functions are standard and should be understood by computers that use BASIC.

Select number 1 from the menu and the screen will ask if you intend applying the same closeness factor to all tasks. Type **N** because Table 4-2 shows a different factor for each task. The screen will ask about confidence factors, and you should again type **N**.

Then the computer will ask how many times task number 1 was observed. Type **56**. It will ask for the closeness and confidence factors, and then it will ask the same questions about task number 2, and so forth. When it asks the number of observations for task number 11, type **END** and the program will return to the menu.

Select number 5; the computer will take just a second to calculate all the ns and display the menu again. This time, select number 6; the results will be displayed as shown in Table 4-3.

Notice that small closeness factors and large confidence factors require larger numbers of observations.

Figure 4-1. Computer program combining closeness and confidence factors.

```
10 '*********************************************************************
20 '** WMEAS.BAS  Calculates number of measurements required in  **
30 '** worker measurement.  For Chapter 4 of Handbook of        **
40 '** Manufacturing and Production Management Formulas,         **
50 '** Charts, and Tables                                        **
60 '*********************************************************************
100 CLEAR,,,32768!: SCREEN 6: COLOR 3,1
110 DIM OBS(200),CLFAC(200),CONFAC(200),NRQD(200)
120 UI$="###": UF$="##.##": UR$="#####"
130 CLS
140 GOSUB 1010                    'Display menu  <<<<<<<<<<<<<
150 CLS
160 ON SELECT GOSUB 1310,1320,1810,2010,2210,2410
170 GOTO 130
180 END
1000 'Subroutine to display menu  $$$$$$$$$$$$$$$$$$$$$$$$
1010 PRINT
1020 PRINT TAB(12)"DATA INPUT/OUTPUT"
1030 PRINT " 1 Enter all new data from keyboard"
1040 PRINT " 2 Add keyboard entries to data in memory"
1050 PRINT " 3 Retrieve data from disk"
1060 PRINT " 4 Save data now in memory"
1070 PRINT
1080 PRINT TAB(12)"WORK WITH DATA IN MEMORY"
1090 PRINT " 5 Calculate observation requirements"
1100 PRINT " 6 Display table of data and calculations"
1110 PRINT
1120 INPUT "Enter number of your selection ",SELECT
1130 IF SELECT>0 AND SELECT<7 THEN RETURN
1140 PRINT "The only choices are 1 through 6"
1150 GOTO 1110
1300 '(1) and (2) Subroutines for entering data  1 2 1 2 1 2 1 2 1 2
1310 N=0: TOTALOBS=0                    'Enter here for selection (1)
1320 INPUT "Will the same closeness factor apply to all tasks? (Y/N) ",CFQ$
1330 IF CFQ$="N" THEN 1380
1340 IF CFQ$="Y" THEN 1370
1350 PRINT "Answer with either Y or N please"
1360 GOTO 1320
1370 INPUT "What PERCENTAGE closeness factor applies to all tasks? ",CF
1380 PRINT
1390 INPUT "Will the same confidence level apply to all tasks? (Y/N) ",CLQ$
1400 IF CLQ$="N" THEN 1520
1410 IF CLQ$="Y" THEN 1440
1420 PRINT "Answer with either Y or N please"
1430 GOTO 1380
1440 INPUT "What confidence factor (z) applies to all tasks? ",Z
1450 IF Z>0 AND Z<=3 THEN 1520
1460 PRINT "That z value is not in the table in this book"
1470 INPUT "Are you sure it's right? (Y/N) ",ZQ$
1480 IF ZQ$="N" THEN 1440
1490 IF ZQ$="Y" THEN 1520
1500 PRINT "Answer with either Y or N please"
1510 GOTO 1460
1520 PRINT
1522 PRINT "How many times was task";N+1;"observed? ... or END";
1530 INPUT " ",OBSERVED$
1540 IF OBSERVED$="END" THEN RETURN
1550 N=N+1
1560 OBS(N)=VAL(OBSERVED$)
1570 TOTALOBS=TOTALOBS+OBS(N)
1580 IF CFQ$="Y" THEN 1600
1590 INPUT "Enter closeness factor as a percentage ",CLFAC(N)
1600 IF CLQ$="Y" THEN 1620
```

Figure 4-1. (continued)

```
1610 INPUT "Enter confidence factor (z) ",CONFAC(N)
1620 PRINT
1630 GOTO 1522
1800 '(3) Subroutine to load data from disk   33333333333333333
1810 INPUT "Name of file on disk ",N$
1820 OPEN "I",#1,N$
1830 N=0: TOTALOBS=0
1840 IF EOF(1) THEN 1890
1850 N=N+1
1860 INPUT #1, OBS(N)
1870 TOTALOBS=TOTALOBS+OBS(N)
1880 GOTO 1840
1890 CFQ$="Y"
1900 CLQ$="Y"
1910 INPUT "What PERCENTAGE closeness factor applies to all tasks? ",CF
1920 INPUT "What confidence factor (z) applies to all tasks? ",Z
1930 CLOSE
1940 RETURN
2000 '(4) Subroutine to save data on disk   444444444444444444
2010 INPUT "Name to be used for saving data ",N$
2020 OPEN "O",#2,N$
2030 FOR C=1 TO N
2040   PRINT #2, OBS(C)
2050 NEXT C
2060 CLOSE
2070 RETURN
2200 '(5) Subroutine to calculate observation requirements  55555555555
2210 FOR C=1 TO N
2220   IF CFQ$="Y" THEN FORMCF=CF/100 ELSE FORMCF=CLFAC(C)/100
2230   IF CLQ$="Y" THEN FORMCON=Z ELSE FORMCON=CONFAC(C)
2240   R=OBS(C)/TOTALOBS
2250   NRQD(C)=(FORMCON*FORMCON*(1-R))/(FORMCF*FORMCF*R)
2260 NEXT C
2270 RETURN
2400 '(6) Subroutine to display full table  6666666666666
2410 PRINT " Task     Number of times    Closeness    Confidence    Number of"
2420 PRINT "number       observed          factor        factor     observations"
2430 PRINT STRING$(64,"-")
2440 ITEMSLEFT=N: START=1: DISPEND=0
2450 IF ITEMSLEFT>20 THEN LINESNOW=20 GOTO 2470
2460 LINESNOW=ITEMSLEFT: DISPEND=1
2470 FOR C=START TO START+LINESNOW-1
2480   PRINT TAB(2) USING UI$;C;
2490   PRINT TAB(15) USING UI$;OBS(C);
2500   IF CFQ$="Y" THEN PRINT TAB(30) USING UF$;CF;: GOTO 2520
2510   PRINT TAB(30) USING UF$;CLFAC(C);
2520   IF CLQ$="Y" THEN PRINT TAB(42) USING UF$;Z;: GOTO 2540
2530   PRINT TAB(42) USING UF$; CONFAC(C);
2540   PRINT TAB(56) USING UR$; NRQD(C)
2550 NEXT C
2560 IF DISPEND=1 THEN 2620
2570 START=START+20
2580 ITEMSLEFT=ITEMSLEFT-20
2590 PRINT
2600 INPUT "Press ENTER for next screen ",E$
2610 CLS: GOTO 2410
2620 INPUT "Press ENTER for main menu ",E$
2630 RETURN
```

Table 4-2. Observation record for computer analysis.

Activity observations	Closeness factor	Confidence factor
56	20	1.98
75	15	2.12
49	25	1.85
83	10	2.25
60	24	1.95
72	25	1.90
38	15	2.50
58	28	1.75
77	30	1.65
78	18	2.00

Choice 4 from the menu—save the data on disk—saves only the number of observations for each task. It is not worth consuming disk space to save the two factors because it is not likely there will ever be the variety of numbers shown in this example.

STOPWATCH METHOD: THE TRADITIONAL WAY FOR TIMING A WORKER

This is the traditional method, whereby an observer watches a worker, timing an entire operation. Each phase of the operation is usually timed while the worker repeats the operation many times. Then the observer does the calculations on the raw data. As with the sampling methods, the minimum number of observations required depends on the closeness and confidence levels desired.

The procedure when timing, instead of sampling, is to make a few observa-

Table 4-3. Output of computer program.

Task number	Number of times observed	Closeness factor	Confidence factor	Number of observations
1	56	20.00	1.98	1033
2	75	15.00	2.12	1521
3	49	25.00	1.85	667
4	83	10.00	2.25	3434
5	60	24.00	1.95	645
6	72	25.00	1.90	460
7	38	15.00	2.50	4444
8	58	28.00	1.75	396
9	77	30.00	1.65	224
10	78	18.00	2.00	899

Press ENTER for main menu

tions, then calculate how many observations are required. If more observations are indicated, make them and then calculate the required number again.

$$n = \left(\frac{z}{c}\right)^2 \left[\frac{a\Sigma t^2}{(\Sigma t)^2} - 1\right]$$

where

a = number of observations made so far

z = confidence factor, determined by the procedure explained under BINARY OBSERVATIONS

c = closeness factor, expressed as a decimal

Σt = sum of observed times

Σt^2 = sum of squares of observed times

Example 4.8

One phase in a worker's position consists of picking up a folded box, opening it, taping it, and attaching a label. The times recorded in ten observations are given in Table 4-4. Also included in the table are the square of each time and the sums of the columns, for use in the formula. The observer wants to know, with ninety-five percent confidence, and within five percent, how long this phase of the job takes. We previously noted that the confidence factor for ninety-five percent is 1.96. The decimal equivalent for five percent is .05. These numbers can be substituted directly into the formula:

$$n = \left(\frac{1.96}{.05}\right)^2 \left(\frac{10(21,046)}{452^2} - 1\right)$$
$$= 46.3$$

Table 4-4. Timing data from ten observations.

Observation number	Time, seconds	Time squared
1	46	2116
2	34	1156
3	52	2704
4	51	2601
5	47	2209
6	32	1024
7	51	2601
8	35	1225
9	53	2809
10	51	2601
	452	21046

As ten observations have already been made, thirty-six more should be added and then n should be calculated again.

COMPUTER PROGRAM FOR STOPWATCH METHOD

Figure 4-2 lists a program that will make the calculation and tell you if more observations should be made. The program is arranged so that you can stop entering observed values at any time, have n calculated, and then continue entering observed values.

PERFORMANCE RATING: HOW TO COMPENSATE FOR WORKERS' REACTIONS TO BEING OBSERVED

The subjective part of observing workers involves estimating the worker's reaction to being observed. Among the reasons why some people work faster when being observed are (1) they concentrate, not allowing their attention to wander, (2) they become nervous, causing the release of more energy into the work, and (3) they try harder because they feel pride in having someone see how well they perform. On the other hand, some workers slow down when being observed because (1) they concentrate on each movement, (2) they become nervous, causing a lack of dexterity, and (3) they deliberately slow down in hopes of causing a standard that is easy or will allow high incentive pay.

Therefore, a part of each individual observation should be a *Performance rating*—an estimate of the percentage of normal for that particular worker. Some writings on this subject relate the observed worker's performance to the expected norm of the group. This book takes the position that the times, quantities, and other measurements form comparisons with others in the group; the correction that is needed is for finding the observed worker's own norm.

$$r_p = \left(\frac{t_e}{t_o}\right) 100$$

where

r_p = performance rating in percent

t_o = observed time

t_e = estimated time for worker to finish task if not being observed

Note that performance rating is less than 100 percent if the worker works slower when observed, and over 100 percent if the worker works faster.

Example 4.9

A worker is observed taking fifty-one seconds to complete an assembly. Immediately afterward, the observer sees the worker slide down in the chair, sigh, and relax for a minute. This strengthens the observer's feeling that nervousness

Figure 4-2. Program to calculate number of observations for stopwatch method.

```
10  '*****************************************************************
20  '** STPWTCH.BAS  Calculates required number of observations for  **
30  '** stopwatch method.  For Handbook of Manufacturing and         **
40  '** Production Management Formulas, Charts, and Tables            **
50  '**                                                              **
60  '*****************************************************************
100 CLEAR,,,32768!: SCREEN 6: COLOR 3,3
110 DIM T(500)
120 U$="###.##"
130 CLS
140 GOSUB 1010                  'Display menu
150 CLS
160 ON SELECT GOSUB 1210,1220,1410,1610,1810,2010
170 GOTO 130
180 END
1000 'Subroutine to display menu  <<<<<<<<<<<<<<<<<<
1010 PRINT
1020 PRINT "1 Enter all new data from keyboard"
1030 PRINT "2 Add entries from keyboard to data in memory"
1040 PRINT "3 Retrieve data stored on disk"
1050 PRINT "4 Save data now in memory"
1060 PRINT "5 Display raw data on screen"
1070 PRINT "6 Calculate required number of observations"
1080 PRINT
1090 INPUT "Select by number ",SELECT
1100 IF SELECT>0 AND SELECT<7 THEN RETURN
1110 PRINT
1120 PRINT "Select only 1 through 6"
1130 GOTO 1010
1200 '(1) and (2) Subroutine to enter data from keyboard  1 2 1 2 1 2 1 2
1210 N=0
1220 PRINT "Enter observed time #";N+1;
1230 INPUT ". . . or END ",TM$
1240 IF TM$="END" THEN RETURN
1250 N=N+1
1260 T(N)=VAL(TM$)
1270 PRINT
1280 GOTO 1220
1400 '(3) Subroutine to retrieve data from disk  333333333333333
1410 INPUT "Name of file on disk ",N$
1420 OPEN "I",#1,N$
1430 N=0
1440 IF EOF(1) THEN CLOSE: RETURN
1450 N=N+1
1460 INPUT #1, T(N)
1470 GOTO 1440
1600 '(4) Subroutine to save data in memory  44444444444444
1610 INPUT "Name to be used for saving ",N$
1620 OPEN "O",#2,N$
1630 FOR C=1 TO N
1640    PRINT #2, T(C)
1650 NEXT C
1660 CLOSE: RETURN
1800 '(5) Subroutine to display raw data  555555555555555
1810 IF INT(N/10)=N/10 THEN R=N/10 ELSE R=N/10+1
1820 FOR ROW=1 TO R
1830    FOR COL=1 TO 10
1840       PRINT TAB(8*(COL-1)) USING U$; T(COL+10*(ROW-1));
1850    NEXT COL
1860 PRINT
1870 NEXT ROW
1880 INPUT "Press ENTER to return to menu ",E$
1890 RETURN
```

Figure 4-2. (continued)

```
2000 '(6) Subroutine to calculate required number of observations  66666666
2010 SUM=0: SUMSQ=0
2020 INPUT "Enter closeness factor as a percentage ",CL
2030 INPUT "Enter confidence factor ",Z
2040 FOR C=1 TO N
2050    SUM=SUM+T(C)
2060    SUMSQ=SUMSQ+T(C)*T(C)
2070 NEXT C
2080 NR=(Z/(.01*CL))^2*(N*SUMSQ/(SUM*SUM)-1)
2090 NR=INT(NR+.5)
2100 PRINT
2110 PRINT NR;"observations are required."
2120 PRINT
2130 IF NR>N THEN 2180
2140 PRINT "You have made enough observations."
2150 PRINT
2160 INPUT "Press ENTER to return to menu ",E$
2170 RETURN
2180 PRINT "You should make";NR-N;"more observations."
2190 GOTO 2150
```

was responsible for the worker's fumbling; that the worker would normally finish the assembly in forty-eight seconds.

$$r_p = \left(\frac{48}{51}\right) 100$$

$$= 94.12 \text{ percent}$$

The observer concludes that this worker will normally complete the assembly in 94.12 percent of the observed time.

Finding Normal Time from the Formula

An observed time corrected by a performance rating is called a *normal time*.

$$\text{normal time} = \text{observed time} \left(\frac{\text{performance rating}}{100}\right)$$

Example 4.10

During one observation, a worker did the task in 1.08 minutes, and the observer assigned a performance rating of 105 percent. What normal time should be recorded?

$$\text{normal time} = 1.08 \left(\frac{105}{100}\right)$$

$$= 1.13 \text{ minutes}$$

Actually, this formula is just a rearrangement of the formula for r_p; it is never necessary to use both of them. They are both given because some analysts find it easier to estimate a performance rating and then calculate the normal time. Others prefer to estimate how long the worker would take if not being watched (normal time) and then calculate the performance rating.

Figuring Average Normal Time

After sufficient observations have been recorded, they can be averaged and adjusted for performance rating.

$$\text{Average normal time} = \frac{\left(\begin{array}{c}\text{Sum of}\\\text{observed}\\\text{times}\end{array}\right)\left(\begin{array}{c}\text{Sum of}\\\text{performance}\\\text{ratings}\end{array}\right)}{100 \text{ (number of observations)}^2}$$

Example 4.11

Two hundred and thirty observations were made of an operation and the sum of the observed times is 210.42 minutes. The sum of the 230 estimated performance ratings is 22,609. What is the average normal time?

$$\text{Average normal time} = \frac{210.42(22,609)}{100(230)^2}$$

$$= 0.90 \text{ minute}$$

Standard Time: Making Allowances for Average Normal Time

One more adjustment should be made before considering the figure as a production standard. It is not expected that the pace during one observation could be maintained during an entire workday. Also, if an assembly line is not perfectly balanced or compensated, there will be delays at some stations. Finally, workers need personal time, and there should be a let-down factor in addition to breaks.

standard time = average normal time + allowances

or

$$\text{standard time} = \text{average normal time}\left(1 + \frac{\text{percentage allowance factor}}{100}\right)$$

Two formulas are given because it is sometimes easier for analysts to think in terms of absolute time; other analysts prefer to think of a decimal factor.

Example 4.12

A team of analysts conducts a study that ends with an average normal time of 7.48 minutes for a certain task. They then estimate that, over an entire shift,

Table 4-5. First three rows of Table A-2.

76496	13519	23285	40803	55138	84706
47058	81511	33915	86108	03605	06477
96257	65891	20708	31000	20350	82072
39102	56042	00514	09705	62758	97812
02146	64876	57220	53153	12565	36661
37552	23494	05968	68155	22426	25266

it is reasonable to expect that, for various reasons, standard time will be three percent higher. What will the standard normal time be?

Use the second formula:

$$\text{Std. normal time} = 7.48 \left(1 + \frac{3}{100} \right)$$

$$= 7.70 \text{ minutes}$$

TECHNIQUE FOR ENSURING RANDOM AND UNBIASED OBSERVATIONS

Observers sometimes go to great lengths to ensure that their sample times are truly random. A serious bias can be introduced into the recordings if the observation times follow a pattern. For example, one observer's pattern might be to check each worker in a given sequence, then start over with the first one and repeat the sequence. At the very least, the sequence time and one of the worker's cycle time could have a common multiple, causing some part of the worker's operation to be noted a disproportionate number of times. It is even possible for the observations to be so synchronized that certain parts of the task are always missed.

Random Number Table

One technique to eliminate bias due to synchronized observations is to use a random number table, such as Table A-2 in the appendix. Assign a series of numbers to each worker and then the next number in the table determines which worker will be observed next.

Example 4.13

Use Table A-2 to control the work sampling observations of twenty-seven workers.

Table 4-6. Table for assigning random numbers to individual workers.

Number	1–3	4–6	7–9	10–12	13–15	etc.
Worker	A	B	C	D	E	→

Table 4-7. Observation plan based on random numbers.

Random number	Worker	Task Observed
76	Z	
49	Q	
61	U	
35	M	
19	G	
23	H	
28	J	
54	R	
08	H	
03	A	
55	S	
etc.	—	

Table 4-5 repeats the first three rows of the random number table from the appendix.

Observing twenty-seven workers requires that you use two-digit numbers, so you can think of the digits in the table being bunched 76, 49, 61, 35, 19, 23, and so forth. Each worker can be assigned three of the two-digit numbers and then every number from 01 through 81 will point to one of the workers. See Table 4-6. Assigning three numbers to each worker instead of one is simply a time-saver. If you assigned one number to each worker, every time a number 28 through 99 came up you would have to discard it and pick the next number. That would also exhaust the list of random numbers faster.

The procedure is this: (1) pick the next two-digit random number from Table 4-5 (from Table A-2); (2) find, from Table 4-6, the worker indicated by the random number; (3) enter the worker on the form to be used for observations, as shown in Table 4-7. It is not important to record the random number on the form in Table 4-7. When a number between 82 and 99 comes up, the easiest policy is to discard it and pick the next number.

After using the 4,800 numbers in Table A-2, you can repeat the table, although that would remove some of the randomness. A better way is, when returning to the beginning of the list, interpret the first two two-digit numbers as the row and column at which to continue picking random numbers.

Some books have longer random number tables, and most computers include random number generators of seemingly unlimited length.

CHAPTER 5

Formulas to Help Minimize Inventory Costs and Improve Ordering Procedures

Inventory is a critical function in any manufacturing company. This chapter examines some of the basic inventory procedures, offering formulas on how to figure costs for unplanned shortages, when to replenish your stock, and how to determine the cost for inventory supplied internally.

ANALYZING INVENTORY POLICIES

At one extreme, a company can order in very small quantities, causing large expenses for processing frequent orders, and often running out of inventory. The other extreme ties up cash, may necessitate borrowing, involves storage costs, tempts pilferers, and may lead to losses through spoilage and obsolescence. Certainly there is an inventory policy between these extremes that will minimize inventory costs.

Analyzing inventory policies can have unexpected side benefits, as when calculations for optimum order size reveal the fact that the procedure for stocking is inefficient.

Symbols used

Let's take an overall look at the subject by examining the symbols used in the equations.

c_i = costs of holding one unit in inventory for one year. Included are direct expenses such as shelves, bins, warehouse space, stock clerks, recordkeeping, and interest on borrowed money. There are less direct costs such as income that money could have earned, and a certain percentage of inventory that, for any number of reasons, never becomes a part of the company's product. Some of these costs are easy to identify; others depend on the experience of the person making the examination. The more thorough the examination, the more it becomes necessary to dig into indirect costs.

c_p = purchase costs. Included here are costs for writing and processing purchase requisitions and orders. The actual cost of the product is not included in any of the formulas in this chapter because it would be equal in all of them. However, c_p includes extra amounts that are incurred by certain policies—such as minimums, shipping, or loss of discount due to small quantity. Also included in this item is the cost of placing the product into inventory; care must be taken to avoid counting certain costs here and again as part of c_i. When inventory is manufactured internally, c_p includes start-up costs and paperwork to begin production.

c_s = costs of being short one unit for one year. Perhaps the company has to purchase a more expensive substitute, or the assembly line workers may be paid while they are given a long coffee break. Workers might be sent home, creating morale problems. Customers might go to a competitor; some of them may not return next time. Most of these costs are hard to identify and estimate. Whether some of the estimates are right or wrong might never be proved. Although estimates may be inaccurate, they usually make the results more reliable than if this item had been ignored.

c_t = total cost of inventory policy. This item includes costs of purchasing, holding, shortages, and others.

c_{to} = total cost of optimum inventory policy. This item is the lowest value c_t reaches as other quantities are varied.

i_h = highest amount in inventory—occurs after order is received.

i_t = inventory level that triggers reorder. When inventory is down to this level, the procedure for replenishment should be initiated. The procedure usually begins internally with writing a purchase requisition; in a small organization the entire procedure may consist of writing a purchase order.

n = number of items ordered at one time.

n_o = number of items ordered at one time in optimal inventory policy. This is the number that will lead to total costs of c_{to}.

n_s = number of items held as safety stock.

r_i = rate (units per day) at which you produce units for inventory. This item is used when inventory is supplied by your own production, usually a work-in-process inventory.

r_u = daily usage rate—the number of units that are drawn from inventory every day.

t_d = time, in fraction of a year, between initiating the purchase procedure and the availability of material to be drawn from inventory. This item includes internal delays at both ends of the cycle; time measurement starts when production people say an order should be placed and ends when production people can use the material.

t_o = time, in fraction of a year, between orders or manufacturing runs, when optimal inventory policy is in effect. This time results in total costs of c_{to}.

u = usage in units per time period.

SITUATIONS WHEN COST OF SHORTAGES IS NOT APPLICABLE

Formulas in this group apply when either shortages are not a possibility, or when there is no cost involved in running out of inventory. For example, a local supplier might make frequent deliveries from its own large supply, or the production effort might be shifted, with no cost, away from the area in which an inventory shortage develops. These formulas also apply when an analyst chooses to ignore costs of shortages as insignificant, or when only a fast approximation, without the refinement of shortage costs, is acceptable.

How Total Inventory Costs Are Found

Total cost is the sum of the cost of holding a given quantity of rotating stock (stock that is constantly being added to or removed from inventory, similar to a revolving charge account), plus the cost of a quantity of safety stock (a cushion that you do not normally expect to use), plus the cost of ordering. Let's assume that items are drawn from inventory at a constant rate, and therefore use average values.

$$c_t = \frac{n c_i}{2} + \frac{u c_p}{n} + n_s c_i$$

Example 5.1

A firm that uses 6,156 pipe nipples a year has been ordering 1,000 at a time. They plan their orders so that new stock is received when they still have 150 left (safety stock). Keeping one nipple in inventory for one year costs $0.75. What is the cost of this policy if each purchase order costs $19.50?

$$c_t = \frac{1,000(0.75)}{2} + \frac{6,156(19.50)}{1,000} + 150(0.75)$$

$$= \$607.54$$

How to Minimize Total Inventory Costs by Determining Optimum Order Size

The quantity to order at one time so as to minimize total inventory costs is found from:

$$n_o = \sqrt{\frac{2\, u\, c_p}{c_i}}$$

Example 5.2

What quantity should the firm in the preceding example order each time?

$$n_o = \sqrt{\frac{2(6,156)19.50}{0.75}}$$

$$= 566$$

Because their holding costs are fairly large, this firm should order fewer at a time. At the same time, an alert analyst should question the procedure for holding items in inventory. If c_i can be reduced, a larger optimum order size can be calculated.

What to Do to Figure Cost of Optimum Order Size

If a firm orders the quantity found by the preceding formula, its inventory policy will have the following cost:

$$c_{to} = \sqrt{2\, c_p\, c_i\, u}$$

Example 5.3

What is the least an inventory policy can cost, given the conditions of the preceding examples?

$$c_{to} = \sqrt{2(19.50)\ 0.75\ (6,156)}$$

$$= \$424.34$$

This amount represents a saving of \$70.70 from the original policy of ordering 1,000 at a time.

Fine tuning

At this point in the analysis an investigator should "fine tune," or look for the best practical quantity to order. If one gross of the item is sent in a convenient storage package, the firm would probably order 576 at a time—after using the first formula again to see if the total cost of that policy is close to the lowest possible cost.

Reorder Trigger Point: When Is It Time to Replenish Your Stock?

This formula determines how many of the item should be in inventory when the next order is placed. It assumes ideal conditions—usage is steady and predictable, and new material will be received exactly when expected. Another assumption is that the order is received in the inventory room at the moment the last of the old items is being withdrawn.

$$i_t = t_d\ u$$

When to reorder

Example 5.4

A production department uses 22,000 inspection certificates per year. When should new ones be ordered if the printer will deliver them in one week?

As t_d must be expressed in years, one week must be written as 1/52 in the formula. Actually, any time unit could be used as long as all variables in the formula are adjusted to be consistent.

$$i_t = \frac{1}{52}\ (22,000)$$

$$= 423$$

If new forms are ordered when there are 423 left, and normal conditions prevail, they will arrive as the last one on hand is being used.

Safety margin

A modification of the formula provides a cushion, or safety margin.

$$i_t = t_d\ u + \text{cushion}$$

Example 5.5

The production department uses 1,423 feet of 3/8-inch hexagonal stock each week. It takes three working days from the time the stock clerk requests new stock until it is available for production use. The department wants a one-day cushion so production can continue if delivery is late. At what stock level should the clerk request new material?

The one-day cushion is 1/5 of the weekly usage.

$$\text{cushion} = 1/5(1{,}423)$$

$$= 285$$

All figures this time will be on a weekly basis. Assuming five working days to a week, the three-day lead time is 3/5 of a week.

$$i_t = \frac{3}{5}(1{,}423) + 285$$

$$= 1{,}139$$

If new material is requested when present stock is down to 1,139 feet, the firm expects to receive new stock when they have 285 feet left.

HOW FINDING A SERVICE LEVEL CAN CONTROL USAGE RATE VARIATIONS

Frequently the usage rate varies from day to day, having an average value and a standard deviation. Likewise, the lead time may not be a dependable fixed time. To include these variations in the calculations, you must establish a service level, which can best be defined by an example. If you say the service level is ninety-five percent, it means that the safety stock will be sufficient ninety-five percent of the time. Implied in this definition is a five percent risk (100 minus service level) of running out of stock before new stock is received.

Let's make the reasonable assumption that both the usage rate and the lead time are normally distributed and are independent of each other. The next three formulas will use figures from appendix Table A-1.

Lead time constant, usage varies

If the lead time is constant, but the usage rate varies with a standard deviation of s_u, the following formula tells when to reorder:

$$i_t = t_d r_u + z \sqrt{t_d}\, s_u$$

Example 5.6

A firm draws washers from stock and places them in assembly kits to pack with the product. They can always obtain delivery of washers in three days and they want to maintain a service level of ninety-five percent.

Records show that 99.8 percent of the days they will pack between 476 and 524 washers, so they consider that they need 500 plus or minus 24 each day. As three standard deviations on each side of center of a normal distribution include 99.8 percent of all data, one standard deviation is 24/3, or 8.

What inventory level should trigger an order for more washers?

To evaluate z in the formula, turn to appendix Table A-1. The table relates distance from the normal curve's center, in standard deviations, to area under the curve. As the curve is symmetrical, 0.5 of the area is to the left of center and 0.5 is to the right. We want to find the point that includes 0.95 of all the area under the curve. The fraction 0.95 will be made up of all the area to the left of center (0.5) plus 0.45 to the right of center, leaving .05 to the right of the area. Look through the body of Table A-1 for 0.45; the nearest is 0.4495. Read the first two digits of z in the first column of the row that contains 0.4495. They are 1.6. Read the second decimal place in the top row of the column that contains 0.4495. It is .04. Therefore, $z = 1.64$.

We can now substitute in the formula for reorder trigger point:

$$i_t = 3(500) + 1.64 \sqrt{3}\, 8$$

$$= 1,523$$

Lead time varies, usage constant

If the lead time is expected to vary, but the usage rate is constant during that time, the formula for trigger point becomes

$$i_t = r_u[t_{davg} + z(s_{\text{lead time}})]$$

Example 5.7

Each of twelve inspectors fills in three report forms every day, so the usage rate is a constant thirty-six per day. It usually takes four days to receive new report forms, and standard deviation on this time is one day. If we want to be ninety-nine percent certain of not running out of report forms, how many should be left when we reorder them?

Using the procedure described in the preceding example, we find that $z = 2.33$.

$$i_t = 36[4 + 2.33(1)]$$

$$= 228 \text{ forms}$$

Usage and lead time vary

The third possibility is that neither usage rate nor lead time is constant—both are known only by their averages and standard deviations. The reorder point is then

$$i_t = r_u(t_d) + z\sqrt{t_d(s_u)^2 + r_d^2(s_{\text{lead time}})^2}$$

Example 5.8

Grinding compound usage averages 2.6 pounds per day, with a standard deviation of 0.7 pounds. The supplier's truck makes deliveries twice a day, and they always provide grinding compound in two days with a standard deviation of one-half day. For a ninety-nine percent service level, we should order when down to:

$$i_t = 2.6(2) + 2.33 \sqrt{2(0.7^2) + 2.6^2(0.5^2)}$$

$$= 9 \text{ pounds}$$

COST OF SHORTAGES INCLUDED: FORMULAS THAT HELP ANTICIPATE INVENTORY DEPLETION

The following formulas recognize that inventories do run out. Estimates of the cost to the firm of such shortages are part of each formula.

How to Calculate Optimum Order Size

Quantity to order

This formula calculates the quantity to order each time so as to minimize inventory costs, including the cost of being out of stock.

$$n_0 = \sqrt{\frac{2 \, c_p \, u \, (c_i + c_s)}{c_i \, c_s}}$$

Example 5.9

A firm's analysts estimate that it costs the firm at the rate of $10.00 per year for each lamp that is out of stock. It uses 18,500 lamps each year, and it costs $0.45 to keep one in inventory for one year. How many should be ordered each time if it costs $21.20 to process an order?

$$n_0 = \sqrt{\frac{2(21.20)18,500(0.45 + 10.00)}{0.45(10.00)}}$$

$$= 1,350$$

An investigation of the formula shows that n_0 decreases when either c_i or c_s is made larger. It decreases when either c_p or u is made larger. These facts can help the analysts decide which parts of the situation should be examined to improve total costs.

Cost of Ordering in Optimum Lots for Recent Inventory Calculations

When inventory is ordered in quantities just calculated, this formula gives the lowest total inventory cost.

$$c_{t0} = \sqrt{\frac{2 \, c_i c_p c_s \, u}{c_i + c_s}}$$

Example 5.10

What is the inventory cost of the policy in the preceding example?

$$c_{t0} = \sqrt{\frac{2(0.45)21.20(10.00)18,500}{0.45 + 10.00}}$$

$$= \$581.19$$

Inventory Level After Receiving Order

When an order is received after there has been a shortage, the incoming material is immediately divided, with part of it going directly to manufacturing and part to inventory. Therefore the highest level in inventory is less than the amount ordered, as shown in the following formula.

$$i_h = \sqrt{\frac{2\,c_p\,c_s\,u}{c_i(c_i + c_s)}}$$

Example 5.11

It costs \$23.60 to write and process an order for washers that cost \$0.22 a year to keep in inventory. The firm uses 15,000 a year. Their analysts estimate that one washer out of stock costs the company at the rate of \$3.10 per year. What is the highest level their inventory will reach?

$$i_h = \sqrt{\frac{2(23.60)3.10(15,000)}{0.22(0.22 + 3.10)}}$$

$$= 1,733$$

HOW TO FIGURE COSTS FOR INVENTORY SUPPLIED INTERNALLY

Let's look at a firm that manufactures items for its work-in-process inventory. Those items will be used as components of a product at the same firm. Typically the manufacture-for-inventory phase is not continuous—it operates until a certain level of inventory has accumulated or until it has produced a certain quantity. Then it stops until inventory is down to a predetermined trigger level. It is, of course, necessary that inventory items be produced faster than the next phase will draw them out.

Chapter 14 includes a computer simulation of a manufacture-for-inventory arrangement. That examination includes a start-up delay—the time after the trigger level is reached but before production of the inventory items actually starts.

Total Inventory Costs to Consider

Even if inventory is not delivered from a supplier, there is a paperwork cost (c_p) each time inventory is needed. It could be a set-up cost. The management systems of some companies include issuing regular purchase orders, but processing them is simplified because they remain internal. Some firms obtain external bids, and use their internal source only when it costs less than using outside suppliers.

$$c_t = \frac{c_p\, u}{n} + \frac{c_i n\left(1 - \dfrac{r_u}{r_i}\right)}{2}$$

Example 5.12

An electronics manufacturer makes power supply subassemblies for inventory at the rate of 30 per day. They make 500 and then stop until more power supplies are needed. The main assembly line draws 18 per day from inventory to supply an annual demand of 4,500 complete assemblies. It costs \$27.80 to cut and process start-up instructions. A subassembly held in inventory incurs holding costs at the rate of \$5.75 per year. What is the cost of this policy?

$$c_t = \frac{27.80(4,500)}{500} + \frac{5.75(500)\left(1 - \dfrac{18}{30}\right)}{2}$$

$$= \$825.20$$

Optimum Batch Size Needed to Ensure Minimal Costs

This formula finds the number of units that should be made each time the subassembly operation is started, to minimize inventory costs.

$$n_0 = \sqrt{\frac{2\,c_p\, u}{c_i\left(1 - \dfrac{r_u}{r_i}\right)}}$$

The general guide for changes is that increasing c_p, u, and r_u make the optimal batch size larger. Increasing c_i or r_i makes n_0 smaller.

Example 5.13

What is the best quantity to make, for minimizing inventory costs, each time the subassembly operation in the preceding example is started?

$$n_0 = \sqrt{\frac{2(27.80)4,500}{5.75\left(1 - \dfrac{18}{30}\right)}}$$

$$= 330$$

How to Avoid Suboptimization by Determining Minimum Inventory Cost

Lowest inventory cost

The lowest inventory cost, when items are made for inventory in batch sizes found from the preceding formula, is given by the following.

$$c_{t0} = \sqrt{2\, c_p c_i\, u \left(1 - \frac{r_u}{r_i} \right)}$$

You could also find c_{t0} by substituting n_0 in the equation for c_t.

Example 5.14

What is the cost of producing the optimum batch sizes from the preceding example?

$$c_{t0} = \sqrt{2(27.80)5.75(4{,}500)\left(1 - \frac{18}{30} \right)}$$

$$= \$758.59$$

By changing from production runs of 500 to 330, the firm saves \$66.61 in inventory costs. This calculation is especially important because it helps the analyst avoid suboptimization. It is possible that production runs of 500 were optimal for reasons other than inventory. Before changing the size of production runs, a comparison should be made to see if the saving in inventory costs more than offsets the other loss.

Calculating Appropriate Time Between Production Runs

When the optimal policy is followed, the length of time between starts of production runs is simply the number produced each time divided by the number used annually.

$$t_0 = \frac{n_0}{u}$$

It is often more convenient to combine the calculation for n_0 into the formula. Then the formula for time between production runs is

$$t_0 = \sqrt{\frac{2\, c_p}{c_i u \left(1 - \dfrac{r_u}{r_i} \right)}}$$

As the usage rate has moved from numerator to denominator (compared to the formula for n_0), an increase in its value shortens the time between production runs.

Example 5.15

If the optimum batch size previously calculated is used, how often will the firm start up its subassembly operation?

$$t_0 = \sqrt{\dfrac{2(27.80)}{5.75(4{,}500)\left(1 - \dfrac{18}{30}\right)}}$$

$$= .0733 \text{ year}$$

Because t_0 is defined as a fraction of a year, it can be converted to days by multiplying it by the number of days per year. Using 250 working days per year allows for holidays but not for a vacation shutdown.

$$t_0 = .0733(250)$$

$$= 18 \text{ working days}$$

To determine if this is truly an optimal policy, an analyst should question how the subassembly workers and machines will be used between production runs. Producing 330 units per day means they operate for eleven working days, and there will be seven working days before another production run starts.

JUST IN TIME ORDERING

The growing use of air freight and rapid computer processing have led some companies to institute systems for keeping inventory levels that approach zero. In the ideal arrangement, suppliers deliver material just when you are ready to use it in manufacturing, and customers take your finished products just as you finish them. One company has saved over $110 million in incoming, work in progress, and finished goods inventories.

A JIT (Just in Time) program is demand-based, and its effectiveness is a function of your ability to forecast demand for your products. With an accurate forecast of your customers' needs, you can avoid finished goods and work in progress inventories. From there you can work backwards to inform suppliers when to deliver materials so you will not need a raw materials inventory. To the extent that you must provide for a statistical uncertainty in your demand forecast, the JIT program will depart from ideal. That is, when you are unable to make an exact forecast, and must produce a safety quantity to be certain of having enough final product, the JIT program cannot achieve its theoretical optimum saving.

Another determinant in the effectiveness of JIT is reliability of incoming materials. If you can depend on the material being 100 percent acceptable, then you can arrange for delivery of exactly the quantity you will need. But if the only way you can be sure of having enough acceptable items is to order extra (the chapter on business statistics will show that three standard deviations extra will give you over ninety-nine percent certainty of not being short), then you will accumulate an inventory of incoming material.

Suppliers can provide your manufacturing needs from one of three bases:

1. From a continuous run. If you purchase an item that is manufactured continuously, the supplier simply fills your orders as you need the item. You will probably still receive a quantity discount for the total you purchase, although you can expect that shipping costs will be higher than if you accepted a large quantity and kept an inventory of incoming materials.

2. From the supplier's inventory. The supplier's optimum arrangement might be to periodically produce a large quantity of the item for you, maintain an inventory of finished goods, and ship to you from that inventory. In effect, you are passing your inventory costs back to the supplier, and the supplier's price will have to reflect that cost plus the cost of set-ups when his inventory needs replenishment.

3. From supplier's special production. If the supplier determines that its costs are lower when it sets up production for each of your orders, instead of maintaining an inventory of finished goods, the supplier will need sufficient lead time from you.

Regardless of the manufacturing arrangement, effective JIT programs result from cooperation between suppliers and customers. Customers make accurate forecasts, give suppliers sufficient lead time, and accept the costs of passing inventory expenses back to suppliers. Suppliers commit themselves to delivery of acceptable items at the time customers need them. Some manufacturers periodically hold get-togethers or symposia with their suppliers to exchange ideas and experiences.

Internal cooperation is as important to you as cooperation with your supplier. Companies with successful JIT programs usually establish engineering-purchasing-inspection-production teams that coordinate the entire program.

The conclusion is that JIT is an alternative inventory policy for some companies. Before adopting it, you should determine the cost of an optimum physical inventory policy, and compare that to the costs of JIT.

COMPUTER PROGRAM FOR INVENTORY USE

Because inventory policy can have a significant effect on profit, it should not be established without extensive investigation. The program listed in Figure 5-1 allows analysts to try many numbers in a short time.

Variable names have been kept to one or two characters so that the program can be copied on any computer that works with standard BASIC. Some computers do not accept **CLS** as a command to clear the screen: line 70 may have to be changed to **HOME**, **CALL CLEAR**, or whatever command is appropriate for the computer being used.

After entering the program, test it with the examples in this chapter. If any

Figure 5-1. Computer program to solve inventory formulas.

```
10 '*********************************************************
20 '** INV.BAS  Inventory programs for Handbook of      **
30 '** Manufacturing and Production Management          **
40 '** Formulas, Charts, and Tables                     **
50 '*********************************************************
60 CLEAR,,,32768!: SCREEN 6: COLOR 3,9
70 CLS
80 PRINT
90 PRINT TAB(20)"COST OF SHORTAGES NOT INCLUDED"
100 PRINT " 1 Total inventory costs"
110 PRINT " 2 Optimum order size and cost"
120 PRINT " 3 Reorder trigger level"
130 PRINT
140 PRINT TAB(20) "COST OF SHORTAGES INCLUDED"
150 PRINT " 4 Optimum order size and cost; highest inventory level"
160 PRINT
170 PRINT TAB(20) "INVENTORY CREATED BY INTERNAL PRODUCTION"
180 PRINT " 5 Total inventory costs"
190 PRINT " 6 Optimum batch size, cost, and timing"
200 PRINT
210 INPUT "Select by number ",SL
220 IF SL>0 AND SL<7 THEN 250
230 PRINT "Select from choices 1 through 6"
240 GOTO 80
250 CLS
260 GOSUB 320                    'Common value inputs
270 ON SL GOSUB 440,500,570,670,780,870
280 PRINT
290 INPUT "Press <Enter> to return to main menu ",E$
300 GOTO 70
310 'Subroutine to input common values  $$$$$$$$$$$$$$$$$$$
320 PRINT "DO NOT USE $ DOLLAR SIGNS"
330 PRINT
340 PRINT "Enter the cost of holding one item in"
350 INPUT "inventory for one year ",CI
360 PRINT
370 PRINT "Enter the paperwork costs of one"
380 INPUT "purchase or internal production order ",CP
390 PRINT
400 INPUT "Enter the number of units used annually ",U
410 PRINT
420 RETURN
430 'Subroutine (1) total inv costs -- no shortages  1111111111111111
440 INPUT "How many units will be ordered at one time? ",N
450 PRINT
460 CT=N*CI/2+U*CP/N
470 PRINT "The cost of this policy is $";CT
480 RETURN
490 'Subroutine (2) optimum order size and cost -- no shortages  2222222222
500 NO=SQR(2*U*CP/CI)
510 CO=SQR(2*CP*CI*U)
520 PRINT "To minimize inventory costs, the firm should"
530 PRINT "order";NO;"units at a time.  This policy"
540 PRINT "will cost $";CO
550 RETURN
560 'Subroutine (3) reorder trigger level -- no shortages  3333333333333
570 PRINT "How many working days between decision to"
580 PRINT "order and availability of material?  Include"
590 PRINT "internal processing time and vendor's time ",D
600 PRINT
```

Figure 5-1. (continued)

```
610 TD=D/250
620 IT=TD*U
630 PRINT "Decision to order should be made when"
640 PRINT "inventory is down to";IT
650 RETURN
660 'Subroutine (4) all formulas recognizing shortages   44444444444444
670 PRINT "Enter the cost of being short"
680 INPUT "one unit for one year ",CS
690 PRINT
700 NO=SQR((2*CP*U*(CI+CS))/(CI*CS))
710 CO=SQR(2*CI*CP*CS*U/(CI+CS))
720 IH=SQR(2*CP*CS*U/(CI*(CI+CS)))
730 PRINT "Minimum inventory costs will be $";CO;"when"
740 PRINT NO;"units are ordered at a time.  Maximum"
750 PRINT "inventory level will be";IH;"units."
760 RETURN
770 'Subroutine (5) total inv costs for internal production   5555555555555
780 GOSUB 980              'Input common values
790 PRINT
800 PRINT "How many units will be made in"
810 INPUT "each production run? ",N
820 CT=CP*U/N+CI*N*(1-RU/RI)/2
830 PRINT "Total inventory costs for this policy"
840 PRINT "will be $";CT
850 RETURN
860 'Subroutine (6) batch size, cost, timing -- internal production  66666666
870 GOSUB 980              'Input common values
880 NO=SQR(2*CP*U/(CI*(1-RU/RI)))
890 CO=SQR(2*CP*CI*U*(1-RU/RI))
900 TO=SQR(2*CP/(CI*U*(1-RU/RI)))
910 WD=TO*250
920 PRINT "Inventory costs will be minimized at $";CO;"when"
930 PRINT NO;"units are manufactured in each production run."
940 PRINT "A new run will start every";TO;"year, or"
950 PRINT "every";WD;"working days."
960 RETURN
970 'Subroutine to input common values for internal production   $$$$$$$$$$$$
980 PRINT "At what daily rate will units be internally"
990 INPUT "manufactured for inventory? ",RI
1000 PRINT
1010 PRINT "At what daily rate will units be"
1020 INPUT "drawn from inventory? ",RU
1030 PRINT
1040 IF RI>RU THEN 1090
1050 PRINT "The system will not function unless inventory is being"
1060 PRINT "supplied faster than it is being drawn out.  Please"
1070 PRINT "enter a realistic set of numbers."
1080 GOTO 980
1090 RETURN
```

answer does not agree with the example's answer, carefully check each line in the appropriate part of the program.

The program is structured with separate subroutines performing each item from the menu. Following is a description of each section of the program, so that any problem can be isolated to a particular area.

Lines 90-250 display the menu and check that the operator's selection is meaningful.

Line 260 calls a subroutine consisting of lines 310-420, regardless of which item is selected from the menu. This subroutine asks for values for c_i, c_p, and u

because they are used in each set of formulas. In addition, some of the other subroutines ask for values for other variables.

After the program completes the subroutine in lines 310-420, line 270 directs it to one of the other subroutines, depending on the operator's choice from the menu.

If either 5 or 6 is selected, values will be needed for r_i and r_u. Therefore the subroutine consisting of lines 970 to 1090 is called by both subroutines—the one comprising lines 770-850 and the one comprising lines 860-960.

A remark line at the beginning of each subroutine identifies the function of that subroutine. Remark lines begin with an apostrophe and are for information only; they have no active part in running the program. Some computers may require that the apostrophe be replaced with the letters **REM** (for remark). Remark lines can be omitted without affecting the running of the program.

Example 5.16

To use the computer for working the first example in this chapter, choose #1 from the menu because the first example calculates *total inventory costs*. Figure 5-2 shows the screen, with the questions the computer asks, the responses (numbers from the first example), and the answer ($495.04).

Figure 5-2.   Screen demonstrating use of computer program to solve first example in Chapter 5.

DO NOT USE $ DOLLAR SIGNS

Enter the cost of holding one item in inventory for one year .75

Enter the paperwork costs of one purchase or internal production order 19.50

Enter the number of units used annually 6156

How many units will be ordered at one time? 1000

The cost of this policy is $495.042

Press ⟨**Enter**⟩ to return to main menu

CHAPTER 6

Quality Control and Its Impact on the Production Process

This chapter looks at quality control and offers formulas for maintaining acceptable production standards. You'll find two types of quality control and how they are used, five key methods for establishing upper and lower control limits, and how to improve your chances for an accurate estimate with two-state probability.

TWO TYPES OF QUALITY CONTROL AND HOW THEY ARE USED

Process control

Specialists in quality control recognize two general types of control. One is *process control*, in which test results are fed back to the production process. Sometimes it is arranged for adjustments to be made continuously; other arrangements do not change the process until tests determine that the system is out of control.

Acceptance testing

Another type of control is *acceptance testing*, as when a customer determines whether a set of goods received meets specifications and should be accepted. This type of testing is also performed by suppliers to determine if a batch of goods should be shipped.

What "Out of Control" Means

The expression means that the process should be checked to determine the reason test results are beyond their limits. Once the problem has been explained (machine out of adjustment, improperly instructed employee), the process is no longer declared out of control.

Two Types of Errors to Consider

Unless 100 percent of the items are tested, quality control is based on statistics and probability. Conclusions are drawn after testing a certain percentage of the items. As long as conclusions are based on samples, *there can be errors*, as seen in election forecasts. If a number between 1 and 1,000 is drawn randomly, you are pretty safe in saying it will be larger than 1. But continue doing this and you can expect to be wrong 1/10 of one percent of the time.

Producer's risk

Two types of errors are possible when conclusions are based on samples. Type I error, also called *producer's risk*, occurs when a batch is wrongly rejected—it actually meets specifications. For example, a customer is promised that every box of 100 items will have no more than two bad ones. An inspector randomly picks ten items from the box and finds two of them bad. As twenty percent of the random sample was bad, it is assumed that twenty percent of the entire box is bad. If the inspector happened to pick the only bad items in the box, this would be a type I error.

Consumer's risk

Type II error, called *consumer's risk*, occurs when a batch that does not meet specifications is passed. If the inspector had unwittingly selected ten good items, even though there were twenty bad ones in the box, there would be a type II error.

HOW TO ESTABLISH UPPER AND LOWER CONTROL LIMITS: FIVE KEY METHODS

Upper control limits (UCL) and lower control limits (LCL) can be established in several ways. Some organizations set limits by rules that are meaningful for specific products. Other rules have general application. Regardless of the source of the limits, they are bounds such that items (or lots) are acceptable if their test results are within LCL and UCL.

1. Process control testing

These formulas are used when a production department tests samples of size n to determine if the process is out of control. The first formula finds how many items should be included in the sample and the second finds UCL.

Items needed for testing

$$n = \left[\frac{z_I\sqrt{p_1(1 - p_1)} + z_{II}\sqrt{p_2(1 - p_2)}}{p_2 - p_1} \right]^2$$

Upper control limits

$$UCL = p_1 + z_I \sqrt{\frac{p_1(1 - p_1)}{n}}$$

where

n = the number that should be tested

p_1 = the portion of rejects normally expected

p_2 = a portion of rejects selected to establish values for the formulas

z_I = number of standard deviations for the probability of a type I error during a batch represented by p_1

z_{II} = number of standard deviations for the probability of a type II error during a batch represented by p_2

The z values are obtained from the table of area under a normal curve in the appendix. For example, setting UCL so there is a four percent probability of error (either type) means that ninety-six percent of the test results will be below the UCL. Table A-1 shows that 0.9599 (the closest in the table to ninety-six percent, which is 0.96) is 1.75 standard deviations from center.

Example 6.1

The manager of a factory that produces large numbers of a small item considers one percent defects to be normal and, during normal operation, wants a twenty percent probability that a good batch could fail the outgoing test. To establish a second reference point, the manager decides that if defects drift to two percent, there should be a ten percent probability that a bad batch could pass the outgoing test.

For the calculations, $p_1 = .01$ and $p_2 = .02$. From Table A-1, $z_I = 0.84$ and $z_{II} = 1.28$.

$$n = \left[\frac{0.84 \sqrt{.01(1 - .01)} + 1.28 \sqrt{.02(1 - .02)}}{.02 - .01} \right]^2$$

$$= 477,481$$

$$UCL = .01 + 0.84 \sqrt{\frac{.01(1 - .01)}{477,481}}$$

$$= .0101$$

The interpretation is that tests should be conducted on 477,481 units. When more than 1.01 percent are defective, the process is out of control. The large number to be tested seems reasonable because of the tight requirements.

2. Mean or $\bar{x}$ criterion

The mean value

This limit applies to the mean value of a sample taken from a lot. The inspector determines the mean value of sample 1 and calls it $\bar{x}_1$. The mean of sample 2 is $\bar{x}_2$, and so forth. After a sufficient number of samples have been measured, the inspector calculates the mean value of all the means, called $\bar{\bar{x}}$.

$$\bar{\bar{x}} = \frac{\bar{x}_1 + \bar{x}_2 + \cdots + \bar{x}_n}{n}$$

The mean of the ranges

The inspector also notes the range of each sample—the difference between the highest and the lowest value in each sample. He or she then calculates the mean of the ranges.

$$\bar{r} = \frac{r_1 + r_2 + \cdots r_n}{n}$$

Next the inspector obtains a factor A, from Figure 6-1. This factor is a function of the number of units in each sample. Then the control limits are

$$LCL = \bar{\bar{x}} - A\bar{r}$$

Figure 6-1. Chart for reading factor _A_ for $\bar{x}$ criterion.

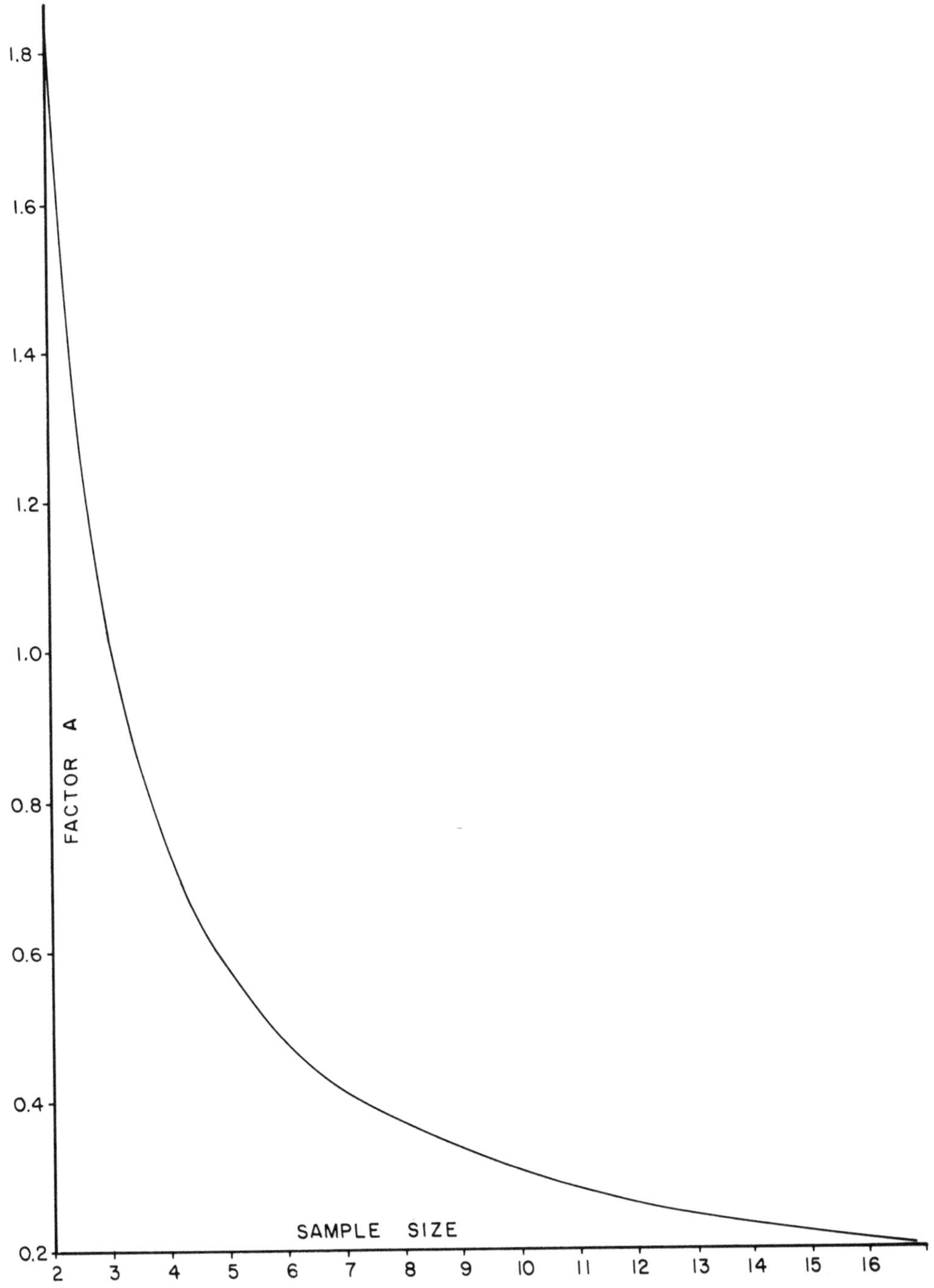

and

$$UCL = \bar{\bar{x}} + A\bar{r}$$

Example 6.2

Resistors for electronic circuits are checked by taking a random sample of 8 out of every 100. You will establish control limits with samples from the first 16 batches. Their test results (resistances in ohms) are shown in Table 6-1.

To calculate $\bar{\bar{x}}$ (the mean of the means), add up all the means and divide by the number of means (16).

$$\bar{\bar{x}} = \frac{66.78 + 68.35 + \cdots + 68.48}{16}$$

$$= 67.93 \text{ ohms}$$

Then the mean of the ranges is the sum of the ranges divided by 16.

$$\bar{r} = \frac{4.85 + 4.88 + \cdots + 6.04}{16}$$

$$= 5.12 \text{ ohms}$$

The final quantity needed for the calculations is A, read from Figure 6-1. It shows that, for samples of 8, A is 0.37.

Now you can calculate the control limits

$$LCL = 67.93 - 0.37(5.12)$$

$$= 66.04 \text{ ohms}$$

Table 6-1. Test results from sixteen samples of eight resistors.

Sample data								Low	High	Range	Mean
69.68	64.94	65.40	64.83	69.31	67.63	67.61	64.84	64.83	69.68	4.85	66.78
68.96	65.79	65.40	69.28	68.72	69.79	70.28	68.61	65.40	70.28	4.88	68.35
65.18	65.31	68.69	68.10	68.45	70.15	67.86	68.46	65.18	70.15	4.97	67.78
67.23	64.70	70.90	65.79	66.41	65.31	66.27	66.10	64.70	70.90	6.20	66.59
69.18	65.42	66.78	69.93	68.59	66.12	71.10	65.67	65.42	71.10	5.68	67.85
70.38	64.96	68.56	66.76	64.71	68.45	70.76	68.95	64.71	70.76	6.05	67.94
65.90	70.40	66.53	67.74	68.45	66.50	70.78	66.65	65.90	70.78	4.88	67.87
67.76	67.26	67.74	66.54	64.98	68.94	68.95	64.60	64.60	68.95	4.35	67.10
65.67	71.23	71.38	71.23	64.82	67.74	70.41	67.87	64.82	71.38	6.56	68.79
70.89	67.87	70.92	68.45	65.94	66.13	68.60	67.48	65.94	70.92	4.98	68.29
68.21	66.51	71.10	65.55	67.63	68.73	67.59	68.95	65.55	71.10	5.55	68.03
65.79	65.40	70.28	68.72	70.79	70.28	68.60	65.19	65.19	70.79	5.60	68.13
68.69	68.10	68.44	70.14	67.86	68.47	69.71	69.68	67.86	70.14	2.28	68.89
65.33	70.52	64.80	70.14	68.57	67.88	67.97	67.25	64.80	70.52	5.72	67.81
67.77	68.69	66.16	67.11	69.21	69.17	68.32	69.50	66.16	69.50	3.34	68.24
70.50	69.94	71.00	68.33	68.61	64.96	65.81	68.68	64.96	71.00	6.04	68.48

and

$$UCL = 67.93 + 0.37(5.12)$$

$$= 69.82 \text{ ohms}$$

When control limits are based on means, you reject any batch from which you draw a sample whose mean is less than 66.04 or more than 69.82 ohms.

3. Range or R criterion

Sample range

Another popular way to set upper and lower control limits is to base them on sample range. The procedure is to first calculate the mean range in the same way as was done for the $\bar{x}$ criterion. Then read factors B and C from Figure 6-2. The formulas for calculating limits are

$$LCL = C\bar{r}$$

$$UCL = B\bar{r}$$

Figure 6-2. Chart for reading factors *B* and *C* for range criterion.

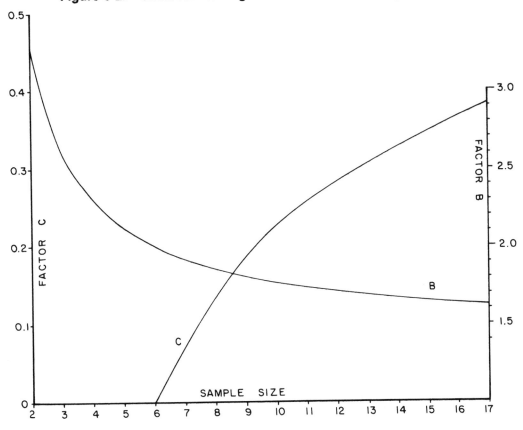

Example 6.3

The resistor manufacturer of the preceding example wants to base acceptance on sample range instead of mean. Calculate the upper and lower control limits.

First read B and C directly from Figure 6-2. They are 1.86 and 0.135 respectively. Then substitute the mean of the ranges, determined in the last example, in the control limit formulas.

$$LCL = 0.135(5.12)$$
$$= 0.69$$
$$UCL = 1.86(5.12)$$
$$= 9.52$$

Any sample of eight resistors whose range of individual test values falls outside those limits will be rejected.

4. Proportion unacceptable or p criterion

This test is often used as a basis for declaring that a product system is functioning normally or is out of control. It uses binary (two-state) data such as pass-fail, present-absent, operating-shut down, and so forth.

The formula uses p, which is defined as the proportion unacceptable in each sample. If sample number one consisted of 25 units and one was unacceptable, then p_1 is 1/25 or .04. This criterion allows the size of the sample to vary slightly (over a range of three or four percent). If the sample sizes vary over a larger range, see the next section. The mean value of p is first calculated in the same way as the other means.

$$\bar{p} = \frac{p_1 + p_2 + p_3 + \cdots + p_n}{n}$$

and then the control limits are

$$LCL = \bar{p} - k \sqrt{\frac{\bar{p}(1 - \bar{p})}{n_s}}$$

$$UCL = \bar{p} + k \sqrt{\frac{\bar{p}(1 - \bar{p})}{n_s}}$$

where

n_s is the sample size. If the samples are not all the same size, n_s is the mean of the sample sizes, but do not use this formula if sample sizes vary by more than a few percentage points.

k is the number of standard deviations to be included in the acceptance band. A value of $k = 3$ is almost universally used, leaving .0013 of normal production outside the band.

Negative number

Whenever the formula for *LCL* evaluates to a negative number, use zero for the lower limit.

Example 6.4

A sample of ten units was taken from each of five typical production batches of 200. The number of rejects in the samples was 2, 1, 3, 1, and 1. How many rejects in one sample would be considered excessive, using three standard deviations as the acceptance band?

$$\bar{p} = \frac{\dfrac{2}{10} + \dfrac{1}{10} + \dfrac{3}{10} + \dfrac{1}{10} + \dfrac{1}{10}}{5}$$

$$= 0.16$$

$$LCL = 0.16 - 3\sqrt{\frac{.16(1 - .16)}{10}}$$

$$= -0.1878$$

Use zero for *LCL*

$$UCL = 0.16 + 3\sqrt{\frac{.16(1 - .16)}{10}}$$

$$= 0.5078$$

The interpretation is that there is no lower number of rejects per sample of ten that would signal that production is not normal. An occasional sample with a *p* of 0.5 (five rejects in a sample of ten) would not necessarily indicate a serious problem. But any time a sample has more than five rejects, the process should be examined closely; it should be declared out of control until the reason for the number of rejects can be identified.

Most quality control specialists would consider five samples an inadequate number on which to base important decisions such as testing policy. These results can be used while more samples are being worked into the formula; then more dependable limits can be established. For example, if the next five samples had 2, 1, 2, 1, and 1 rejects, then *UCL* changes from more than 0.5 to less than 0.5. This change was just enough to change a sample with five rejects into a signal that the process is out of control.

5. When sample size varies over a significant range

The preceding formula is restricted to plans that take samples of nearly uniform size. Several methods have been described in the literature for use when it is more convenient to use samples of differing sizes; a popular one is given here.

Varying size samples

After a series of inspections, $\bar{p}$ is calculated by dividing the total number of defects by the total number of units inspected. Then it is used to find the following constant:

$$k = 3\sqrt{\bar{p}(1 - \bar{p})}$$

and *UCL* and *LCL* are calculated from

$$UCL = \bar{p} + \frac{k}{\sqrt{n}}$$

$$LCL = \bar{p} - \frac{k}{\sqrt{n}}$$

This method calculates limits for each series of inspections. Each series uses the overall average $\bar{p}$ but the specific n for the series.

As n can vary, a series of inspections can be made at fixed periods of time instead of when a fixed number of units is ready. Inspections can be made once a day, once an hour, or sporadically.

Example 6.5

Table 6-2 shows the results of inspections for twenty days. The last four columns show values calculated from the preceding formulas. First $\bar{p}$ is shown (defect ratio), then the control limits are shown, and the last column shows whether the day's testing indicates that the process is in control or out of control. The only day out of control is when the defect ratio is 0.232 and the upper control limit is 0.222.

The sequence of calculations is (1) add the total number inspected for the period, (2) add the total number of defects, (3) divide the second sum by the first to find $\bar{p}$, (4) calculate k from the formula just given, (5) for each day, calculate *UCL* and *LCL*, and (6) for each day, determine whether the process is in or out of control. After the inspection routine has been well established, the long-run defect ratio can be used, allowing current determination of in or out of control.

In this example, *LCL* is set to zero whenever it calculates to a negative number. When counting defects, as in this example, some organizations set the *LCL* for each day to zero, reasoning that the fewer defects the better. Others, after an economic analysis, conclude that a very low defect ratio means too much effort is being put into solving small production deviations; they are below the point of diminishing returns.

When the measurement is other than a count, and a negative *LCL* has meaning, it should retain its calculated value.

Table 6-2. Inspections of varying sample sizes.

Day	Number Inspected	Defective	Defect Ratio	LCL	UCL	Control
1	49	9	0.184	0.000	0.254	in
2	69	9	0.130	0.000	0.232	in
3	87	8	0.092	0.013	0.219	in
4	51	12	0.235	0.000	0.251	in
5	82	19	0.232	0.010	0.222	out
6	66	4	0.061	0.000	0.235	in
7	67	5	0.075	0.000	0.234	in
8	84	5	0.060	0.011	0.221	in
9	78	14	0.179	0.007	0.225	in
10	75	2	0.027	0.005	0.227	in
11	64	1	0.016	0.000	0.236	in
12	54	11	0.204	0.000	0.247	in
13	59	8	0.136	0.000	0.241	in
14	74	11	0.149	0.004	0.228	in
15	56	2	0.036	0.000	0.245	in
16	67	8	0.119	0.000	0.234	in
17	88	2	0.023	0.014	0.219	in
18	85	9	0.106	0.012	0.220	in
19	64	14	0.219	0.000	0.236	in
20	58	7	0.121	0.000	0.242	in

TWO-STATE PROBABILITY: HOW TO IMPROVE YOUR CHANCES FOR AN ACCURATE ESTIMATE

Let's look now at a very useful formula that answers questions such as what is the probability of two rejects in a sample of seven? In addition to the answers it provides directly, two-state (binomial) probability is used as a component of other formulas, as will be shown in a later section.

Factorials and Combinations: A Brief Mathematic Review

If your mathematics skills are current, you can go directly to the formulas. Otherwise, reviewing the next few paragraphs will head off frustrations when you work with the formulas.

Factorial

The factorial of a given number is equal to the product of all integers from one through the given number. Six factorial is 1(2)3(4)5(6), which equals 720. It is written with an exclamation mark—6!.

There are no shortcuts; the entire string must be multiplied. However, if you have previously calculated 5!, you can find 6! by multiplying 5! by 6.

There is no such thing as the factorial of a negative number or of a noninteger. Calculators often have built-in factorial functions, and they give an error signal (flash, spell out *error*, or whatever each model's error signal is) when asked to find factorials that do not exist.

Zero factorial is defined as having a value of one. This fact is derived in many texts, but we treat it here as a definition to be accepted.

Combinations

In its technical sense, taken from statistics, a combination is a specific selection of all or part of a group. For example, five workers (named A, B, C, D, and E) unload trucks of raw material. One drives the forklift, one keeps tally on a clipboard, and three do the lifting. One *combination* of lifters could be A, B, and C. Another could be B, C, and D.

Statisticians refer to n things taken r at a time; in this example there are five things taken three at a time. The formula we will use in a moment will show that we can make ten combinations of five things taken three at a time.

Among the many ways books indicate the number of combinations of n things taken r at a time are

$$^nC^r \text{ or } {}_nC_r \text{ or } \binom{n}{r}$$

It is important to note that rearranging the order in which members of a combination are named does *not* make a new combination. The first selection of workers A, B, and C is the same combination as B, A, and C. In other words, you are not identifying a chief lifter, an assistant lifter, and an ordinary lifter—just three lifters.

The formula for the number of combinations of n things taken r at a time is

$$\binom{n}{r} = \frac{n!}{r!(n-r)!}$$

Example 6.6

The production department has a pool of label makers and one is used by each of four assembly lines. In how many combinations can the four label makers be taken from the pool?

The answer is the evaluation of six things taken four at a time.

$$\binom{6}{4} = \frac{6!}{6!(6-4)!}$$
$$= 15 \text{ ways}$$

Table

When both n and r are 14 or less, Table 6-3 can save you a lot of arithmetic. Simply find n in the left-hand column, then move along that row to where it intersects the column headed by r.

Table 6-3. Table of combinations.

X N	0	1	2	3	4	5	6	7	8	9	10	11	12	13	14
1	1	1													
2	1	2	1												
3	1	3	3	1											
4	1	4	6	4	1										
5	1	5	10	10	5	1									
6	1	6	15	20	15	6	1								
7	1	7	21	35	35	21	7	1							
8	1	8	28	56	70	56	28	8	1						
9	1	9	36	84	126	126	84	36	9	1					
10	1	10	45	120	210	252	210	120	45	10	1				
11	1	11	55	165	330	462	462	330	165	55	11	1			
12	1	12	66	220	495	792	924	792	495	220	66	12	1		
13	1	13	78	286	715	1287	1716	1716	1287	715	286	78	13	1	
14	1	14	91	364	1001	2002	3003	3432	3003	2002	1001	364	91	14	1

Example 6.7

Eight people are trained to operate a certain machine. One person is to be assigned to each of three such machines. In how many ways can the assignments be made?

You must find the number of combinations of eight things taken three at a time. In Table 6-3, find the intersection of the row that starts with $n = 8$ and the column headed by 3. The number of combinations is 56.

Probability defined

One important use of combinations is in calculating probabilities, such as the probability that a certain combination will occur. Probability is defined as

$$P = \frac{\text{The number of ways a particular thing can happen}}{\text{The total number of ways possible}}$$

Example 6.8

We use mixtures of bright colors to communicate liveliness in our products. One item is packaged in a cylindrical can with the top divided radially into three segments, each painted a different color. A painting machine selects three colors at random from ten available. What is the probability of any specific color combination (say red, yellow, green) on any one can?

There is only one combination that is red, yellow, green, so the numerator is one. The denominator is the number of ways the machine can select three colors—the number of combinations of ten things taken three at a time.

$$P = \cfrac{1}{\dbinom{10}{3}}$$

$$= \frac{1}{120}$$

$$= .0083$$

Binomial probability

Applications for this statistic in production are almost limitless. It answers questions such as "if five percent of the output is unacceptable, what is the probability that two out of a random selection of ten will be unacceptable?" We'll calculate that one shortly.

The formula is

$$P(r) = \binom{n}{r} p^r (1 - p)^{n-r}$$

where

$P(r)$ = probability of r occurrences in n tries
p = probability of occurrence in one try

Example 6.9

Let us answer the question in the last paragraph. Because five percent of the output is unacceptable, the probability is .05 that any unit selected at random is unacceptable. Ten units are selected, so $n = 10$. The question asks about finding two bad ones in the group of ten, so $r = 2$.

$$P(2) = \binom{10}{2} (.05)^2 (1 - .05)^{10-2}$$

$$= .0746$$

There is a bit of arithmetic involved, even when a calculator and Table 6-3 are used. Quick answers are often needed, so the appendix includes a table for n from one to twenty.

To use the table, locate n in the first column. Then, in the group associated with that n, locate r. Move to the right along that row to the column headed by p and read the probability at the intersection.

Binomial probabilities tables are included in many statistics and reference books. They always save space by not including p values larger than 0.5, taking advantage of the table's symmetry. The best way to show how to find binomial probabilities when p is larger than 0.5 is with an example.

Example 6.10

In a certain experimental production, sixty percent ($p = 0.6$) of the units fail inspection. If seven units are selected at random, what is the probability that three of them will be rejected?

As 0.6 is not on the table, use $1 - p$, or 0.4. When making that substitution, you must also interchange the number for which you are looking and the number for which you are not looking. That is, instead of looking for three rejects, you will look for seven minus three, or four, acceptable units.

In the first column find seven, then find four in the second column of that group. Move along that row to the column headed 0.4, and the probability at the intersection is 0.1935.

This method is easy to remember if you think that *rejecting* three units out of seven is the same as *not rejecting* (accepting) four units out of seven, and the probability of accepting is one minus the probability of rejecting.

Cumulative probability

We select a sample of five units from a production batch and say that the batch will be rejected if there are two bad units in the sample. But that policy would have us pass batches with three, or even five, bad units in a sample of five. The policy should really call for rejection if two *or more* in the sample of five are bad. That is, if two, three, four, or five are unacceptable. To find the probability that the batch will be rejected, we have to add the separate probabilities

$$P(2 \text{ or more}) = P(2) + P(3) + P(4) + P(5)$$

Example 6.11

A chemical used in production deteriorates on the shelf so that twenty percent of the bottles are not usable after one year. If you select six bottles at random, what is the probability that at least three are unusable?

The probability of at least three is the same as the probability of three or more.

$$P(3 \text{ or more}) = P(3) + P(4) + P(5) + P(6)$$

These probabilities can be read directly from the table in the appendix, in the .20 column.

$$P(3 \text{ or more}) = .0819 + .0154 + .0015 + .0001$$

$$= .0989$$

Some of the arithmetic can be avoided when more than half the numbers in a column are being added. As mentioned at the beginning of the section, these formulas apply when only two things can happen (pass or fail in this example).

As there can be no third condition, the probability of those failing plus the probability of those passing must equal one. Therefore, instead of finding the probability of three or more, you could have found the probability of two or less and subtracted that from one.

$$P(3 \text{ or more}) = 1 - P(2 \text{ or less})$$
$$= 1 - (P(2) + P(1) + P(0))$$
$$= 1 - (.2458 + .3932 + .2621)$$
$$= .0989$$

Another timesaver is the cumulative binomial probability table in the appendix. This table gives cumulative probabilities from $r = 0$. In the last example we could look across from $n = 6$ and $r = 2$ to the intersection with $p = .20$; that number is the sum of $P(2) + P(1) + P(0)$, or .9011. Subtracting that result from one gives .0989, the same as we found when adding the probabilities manually.

AVERAGE OUTGOING QUALITY OF SHIPMENTS

When an organization follows a certain policy, such as replacing the defective units in a sample, it is not difficult to calculate the probable percent of defects in the overall shipment.

Rejects in Sample Replaced

One hundred percent inspection

If the sample does not contain enough defects to call for rejecting the entire batch, the usual procedure is to replace the defects in the sample with known good units, return the sample to the batch, and pass the entire batch. If the entire batch is rejected, it is given 100 percent inspection. That policy will result in the following outgoing quality

$$AOQ = \frac{P(\text{accept})p(n_b - n_s)}{n_b}$$

where

AOQ = average outgoing quality—the fraction of defects in outgoing batches

$P(\text{accept})$ = probability that any batch will be accepted

p = fraction of defects in batches going to inspection

n_b = number in batch

n_s = number in sample

Example 6.12

Machines pack their output in boxes with four rows of forty units each. Records show that there are generally two percent defects in a box. Inspectors take eight units at random from each box for a thorough test. If they find no more than one bad unit they exchange it for a good one, return all eight, and pass the box. If there is more than one bad unit, they test all 160 in the box and replace every bad unit with a known good one. What is the average outgoing quality?

The first step is to calculate the probability that a box will be passed by the inspectors. That is, the probability that a box has zero or one defect.

$$P(\text{accept}) = P(0) + P(1)$$

$$= \binom{8}{0}(.02)^0(.98)^8 + \binom{8}{1}(.02)^1(.98)^7$$

$$= 0.9897$$

Then average outgoing quality is

$$AOQ = \frac{0.9897(.02)(160 - 8)}{160}$$

$$= .0188$$

$$= 1.88 \text{ percent defects}$$

Suppose that isn't sufficient improvement over the two percent average defects the firm would ship if it shipped machine output without any inspection? The firm could change its policy to pass only boxes with zero defects in the sample. That would improve AOQ to 1.62 percent.

For further improvement the firm can increase the sample size to sixteen, and still pass only boxes with no defects in the sample.

$$P(\text{accept}) = 0.7238$$

$$AOQ = \frac{0.7238(.02)(160 - 16)}{160}$$

$$= 1.30 \text{ percent}$$

There can be many combinations of sample size and acceptance criterion, each resulting in a value of average outgoing quality. Table 6-4 shows the results for up to thirty in a sample, and up to eight defects as the cutoff for acceptance. The three results just calculated—1.88, 1.62, and 1.30—could have been found from this table.

Control of *p*

These attempts to improve outgoing quality have not experimented with the percent defects produced by the manufacturing process. It often happens

Figure 6-3. Average outgoing quality as a function of initial fraction defective.

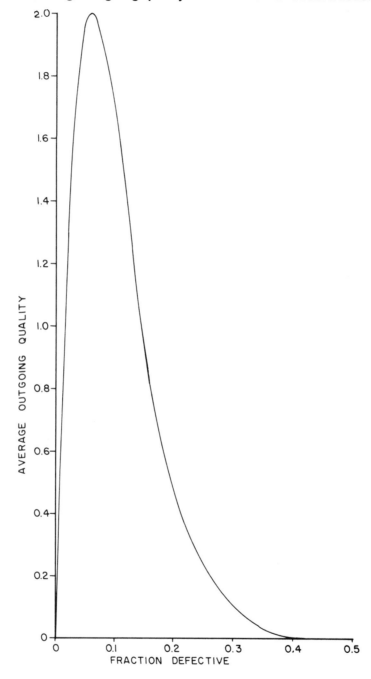

that improvements in the process are not possible, or could be effected only at a cost far larger than can be justified.

However, there are situations where p is controllable; it is worthwhile to examine the way it affects AOQ. If p is very small, boxes will be shipped with very few defects. If p is very large, many batches will undergo 100 percent inspection and have all their defective units replaced. As either a very low or a very high value of p results in a small number of average defects after inspection, it is reasonable to conclude that some intermediate value of p results in a maximum percentage of outgoing defects.

This conclusion is verified by Figure 6-3, which shows AOQ peaks at about two percent when p is about .06. Readers who have maintained their calculus skills can find that the exact maximum is 2.01 percent, when $p = .0588$.

The highest point AOQ reaches is called average outgoing quality limit (*AOQL*). It is the largest percent of average rejects that could result from a given policy.

Table 6-4. AOQ as a function of sample size and acceptance criterion.

Sample Size	Acceptance criterion								
	0	1	2	3	4	5	6	7	8
1	1.95	1.99							
2	1.90	1.97	1.97						
3	1.85	1.96	1.96	1.96					
4	1.80	1.95	1.95	1.95	1.95				
5	1.75	1.93	1.94	1.94	1.94	1.94			
6	1.71	1.91	1.92	1.92	1.93	1.93	1.93		
7	1.66	1.90	1.91	1.91	1.91	1.91	1.91	1.91	
8	1.62	1.88	1.90	1.90	1.90	1.90	1.90	1.90	1.90
9	1.57	1.86	1.89	1.89	1.89	1.89	1.89	1.89	1.89
10	1.53	1.84	1.87	1.87	1.87	1.88	1.88	1.88	1.88
11	1.49	1.83	1.86	1.86	1.86	1.86	1.86	1.86	1.86
12	1.45	1.81	1.85	1.85	1.85	1.85	1.85	1.85	1.85
13	1.41	1.79	1.83	1.84	1.84	1.84	1.84	1.84	1.84
14	1.38	1.77	1.82	1.82	1.82	1.83	1.83	1.83	1.83
15	1.34	1.75	1.81	1.81	1.81	1.81	1.81	1.81	1.81
16	1.30	1.73	1.79	1.80	1.80	1.80	1.80	1.80	1.80
17	1.27	1.71	1.78	1.79	1.79	1.79	1.79	1.79	1.79
18	1.23	1.69	1.77	1.77	1.77	1.77	1.78	1.78	1.78
19	1.20	1.67	1.75	1.76	1.76	1.76	1.76	1.76	1.76
20	1.17	1.65	1.74	1.75	1.75	1.75	1.75	1.75	1.75
21	1.14	1.62	1.72	1.74	1.74	1.74	1.74	1.74	1.74
22	1.11	1.60	1.71	1.72	1.72	1.72	1.73	1.73	1.73
23	1.08	1.58	1.69	1.71	1.71	1.71	1.71	1.71	1.71
24	1.05	1.56	1.68	1.70	1.70	1.70	1.70	1.70	1.70
25	1.02	1.54	1.67	1.69	1.69	1.69	1.69	1.69	1.69
26	0.99	1.52	1.65	1.67	1.67	1.67	1.67	1.68	1.68
27	0.96	1.49	1.64	1.66	1.66	1.66	1.66	1.66	1.66
28	0.94	1.47	1.62	1.65	1.65	1.65	1.65	1.65	1.65
29	0.91	1.45	1.61	1.63	1.64	1.64	1.64	1.64	1.64
30	0.89	1.43	1.59	1.62	1.62	1.62	1.62	1.63	1.63

Money Flow and Interest: How to Calculate Your Earning Power

This chapter gives you formulas for calculating the earning power of money. It starts with formulas for simple interest, then goes on to compound interest, present value, stream of payments, and a payback table.

For example, in the "Simple Interest" section you'll find a basic formula for calculating interest earnings, as well as how to determine initial principal, how to calculate the interest rate when you know other quantities, and how to reach your investment goal by determining the amount you need to invest. The "Compound Interest" section contains formulas for single payments and payments to be made at regular intervals. Annual and monthly interest compounding is also explored.

Computer programs are included to help with the formulas because practical applications usually require repeated calculations or experimentation with different numbers.

HOW TO USE THE FORMULAS IN THIS CHAPTER

The formulas in this chapter also apply to earnings by names other than interest, such as "dividends." As long as the earnings are related by ratio to a principal, the formulas can be used.

Although interest rates are usually expressed as percentages, they are used as decimals in calculations. To convert a percentage figure, remember that *per* means to divide, and *cent* refers to 100. Therefore, 17 percent means 17 divided by 100, or 0.17. Some prefer to remember to move the decimal point two places to the left before using a percent figure in a calculation.

In the same way, when an interest rate is calculated in a formula, the answer will be a decimal; it can be converted to a percent figure by moving the decimal point two places to the right.

Symbols used

a = amount in fund, including original principal, earnings, and other contributions

c = number of compounding periods per year

e = amount earned through interest

m = number of times original principal will multiply

p = principal, or original amount

r = annual rate of interest, in decimal form

s = payment made to increase fund or reduce debt, at end of compounding period

t = time—number of years that interest is earned

v_p = present value of money

Simple Interest

The formulas in this section apply when the principal on which the interest is earned remains unchanged. By contrast, interest is compounded when the

106

amount earned in one period is added to the principal before calculating earnings for the next period.

Most corporate bonds are examples of simple interest. If a $1,000 bond pays twelve percent, the debtor pays $120.00 to the bondholder every year. As the $120.00 is paid out instead of being added to the principal, the debt on which interest is calculated remains at $1,000.

INTEREST EARNINGS: TWO MAIN TYPES

The basic formula

The amount that is earned through interest is found from

$$e = prt$$

When the length of time is given in days, divide by the number of days in a year to get t for the formula. It is acceptable for most business transactions to use 360 days; the result is called *ordinary* simple interest. Using 365 or 366 days yields *exact* simple interest. When time is given in months, divide by twelve.

Example 7.1

Your production department buys plating equipment for $67,000. The seller will wait three months for payment, charging at an annual rate of 13.6 percent. How much will it cost the department to accept this credit?

$$e = 67,000(0.136)(3/12)$$
$$= \$2,278$$

How to Determine Initial Principal

The formula for simple interest earnings can be rearranged for determining the initial principal when a given amount must be earned.

Initial amount

$$p = \frac{e}{rt}$$

Example 7.2

Production department employees have been promised the interest on a "party endowment" and the first party will be in fifteen months. How much should be put into the endowment if it pays 9.5 percent interest and $450 will be needed?

$$p = \frac{450}{0.095\left(\dfrac{15}{12}\right)}$$
$$= \$3,789.45$$

How to Calculate Interest Rate When You Know Other Quantities

This formula is another rearrangement of the first one—this time for convenience in calculating the interest rate when other quantities are known.

Rate unknown

$$r = \frac{e}{pt}$$

Example 7.3

If an investment of $10,000 is to earn $700.00 in ten months, at what rate of interest must it earn?

$$r = \frac{700}{10,000 \left(\dfrac{10}{12} \right)}$$

$$= .084$$

The interest rate must be 8.4 percent.

Example 7.4

A customer wants to place a $56,000 order, but cannot pay cash. Instead, he offers to pay $60,000 in six months. If you accept, what rate of simple interest will the customer be paying?

The difference between $56,000 and $60,000 is the amount you will earn in six months.

$$r = \frac{60,000 - 56,000}{56,000 \left(\dfrac{6}{12} \right)}$$

$$= 0.1429$$

The customer will be paying 14.29 percent interest.

Calculating Time Needed to Earn a Given Amount

Another rearrangement of the formula makes it convenient to calculate how long it will take a certain principal to earn a given amount.

Time calculation

$$t = \frac{e}{pr}$$

Example 7.5

A fund of $17,500 has been marked for keeping the production department's library updated. When the fund earns $1,500, new purchases will be made. How long will it take to earn that amount if the fund is invested at 11.2 percent simple interest?

$$t = \frac{1,500}{17,500(0.112)}$$

$$= 0.77 \text{ year}$$

The decimal part of a year can be converted to months by multiplying it by twelve.

$$t = 0.77(12)$$

$$= 9.24 \text{ months}$$

or about nine months and one week.

How to Find the Total Amount in a Fund

To find how much is in a fund, add the earnings to the principal. However, it is usually easier to use the following formula, which calculates the interest earned and adds it to the principal.

$$a = p(1 + rt)$$

where a = amount in fund after including interest earnings.

Example 7.6

An investment earns simple interest at the rate of 9.5 percent per year. What is the investment worth eleven months after it is started with $16,200?

$$a = 16,200[\, 1 + .095(11/12)]$$

$$= \$17,610.75$$

How Much You Need to Invest to Reach Your Goal

When the total amount equation is rearranged in this form, it is convenient for finding how much must be invested to build up to a given amount.

Principal for total amount

$$p = \frac{a}{1 + rt}$$

Example 7.7

It is necessary to have $83,000 in eighteen months. If an investment pays fourteen percent, how much should be put into it now?

$$p = \frac{83,000}{1 + 0.14\left(\dfrac{18}{12}\right)}$$

$$= \$68,595.04$$

Calculating Time for Total to Reach a Given Amount

The final arrangement of the total amount formula calculates how long it takes to build the fund to a given amount.

$$t = \frac{a - p}{p\,r}$$

Example 7.8

A fund will be used for modernization when it has $40,000. How long will it take if it is started with $25,000, earning 10.7 percent?

$$t = \frac{40,000 - 25,000}{25,000\,(0.107)}$$

$$= 5.61 \text{ years}$$

It will take five years, six and a half months.

How Long It Takes for Principal to Multiply by a Given Factor

This formula is a generalization of the preceding example. It finds how long it takes for a fund to increase by a given factor when the fund is invested at simple interest.

$$t = \frac{m - 1}{r}$$

Example 7.9

How long will it take an investment to triple if it earns 12.6 percent simple interest?

$$t = \frac{3 - 1}{0.126}$$

$$= 15.87 \text{ years}$$

Example 7.10

How long will it take an investment to increase to half again its present value if it is earning 9.8 percent?

Half again means it will increase by fifty percent. It therefore increases by a factor of 1.5.

$$t = \frac{1.5 - 1}{0.098}$$

$$= 5.1 \text{ years}$$

COMPOUND INTEREST

Most business interest calculations involve compound interest because in that case the interest earned in one period is added to the principal before calculating interest for the next period. The calculations become more involved because the principal keeps changing.

For example, if a business borrows $1,000 for twenty-four months at ten percent simple interest, it will owe $200 interest plus the original $1,000 at the end of twenty-four months. But if interest is compounded annually, the $100 interest earned in the first year is added to the principal at the end of one year, and the amount that earns interest during the second year is $1,100. Interest in the second year is $110, so the total earned by annual compounding is $210. If interest had been compounded monthly, periodic earnings would be added to the principal every month.

The shorter the compounding period, the sooner interest earned is added to the principal. Therefore, for a given annual rate, the shorter the compounding period, the higher the total interest earnings.

Unless it is necessary to know the value of the fund after each compounding period, formulas allow you to calculate the final amount directly. That is, for a twenty-four-month loan compounded monthly, you do not have to calculate one month's earnings and add them to the principal twenty-four times. The formulas use exponents and logarithms, making them difficult to solve without tables; a hand calculator makes the job still easier. At the end of this chapter is a computer program that will solve all the formulas.

In general, for a given set of values for time, principal, and annual rate of interest, the more compounding periods per year, the more will be earned.

Single Payment Formulas

The first series of formulas applies to a single payment. That is, except for periodic interest earnings, no additional payments are made to or from the principal.

Calculating the Future Amount of a Principal

This is the standard compound interest formula that calculates the value of a fund, including interest.

The standard formula

$$a = p \left[1 + \frac{r}{c} \right]^{ct}$$

If interest is compounded once per year, $c = 1$ and the formula simplifies to

Interest compounded annually

$$a = p(1 + r)^t$$

Example 7.11

A company lends $22,000 at fourteen percent compounded monthly for eight years. If no payments are made during the eight years, how much will be due?

$$a = 22,000 \left(1 + \frac{0.14}{12} \right)^{8(12)}$$

$$= \$66,991.08$$

Example 7.12

One question that should be asked when analyzing a proposed purchase of capital equipment is how much you would benefit from an alternative use of the money. Suppose a new stamping machine costing $56,000 is being considered. It will last seven years and then is not expected to have any resale value. How much should the machine earn if another use suggested for the money is known to earn twelve percent annually and will preserve the original amount?

The alternative investment will be worth

$$a = 56,000(1 + 0.12)^7$$

$$= \$123,798.16$$

Therefore the new machine should be considered only if it can be depended on to increase profits by that amount during the seven years of its contribution.

How Long It Takes to Earn a Given Amount

This formula is the same as the preceding one, but rearranged so it calculates how long it will take the fund to reach a certain value.

$$t = \frac{\log a - \log p}{c \log \left(1 + \dfrac{r}{c}\right)}$$

If compounding is once a year, the formula simplifies to

$$t = \frac{\log a - \log p}{\log (1 + r)}$$

Although *log* is usually the abbreviation for common logarithms (base 10), any base can be used as long as it is used consistently throughout a calculation. Natural logarithms (base e) are often more convenient on calculators.

Example 7.13

The board of directors has provided $38,000 as the start of a fund to build a new employees' recreation center. When the fund has $100,000, the board will provide any additional amount needed to start construction. If the money earns at the rate of 12.8 percent compounded daily, and no more is added to it, when can construction begin?

$$t = \frac{\log 100,000 - \log 38,000}{360 \log \left(1 + \dfrac{0.128}{360}\right)}$$

$$= 7.56 \text{ years}$$

Calculating the Interest Rate for a Given Amount of Growth

Another rearrangement of the same formula makes it convenient to calculate the interest rate necessary for a given amount of growth.

Exponents

$$r = c \left[\left(\frac{a}{p}\right)^{1/ct} - 1 \right]$$

Roots

$$r = c \left[\sqrt[ct]{\frac{a}{p}} - 1 \right]$$

Two equivalent forms of the equation are shown because some calculators are arranged to work better with fractional exponents and some with fractional roots.

Example 7.14

A company is to borrow \$29,000 for six and a half years and then \$58,000 will be due. If interest is to be compounded quarterly, what rate of interest is being charged?

$$r = 4\left[\left(\frac{58,000}{29,000}\right)^{1/4(6.5)} - 1\right]$$
$$= 0.1081$$

The debtor is paying at the rate of 10.81 percent.

How Long Will It Take to Increase Your Principal by a Given Factor?

A frequent question is, "How long will it take to double my money?" or multiply it by some other factor. This formula answers the question for compound interest.

$$t = \frac{\log m}{c \log\left(1 + \dfrac{r}{c}\right)}$$

Example 7.15

How long will it take to double an investment if it earns 11.2 percent compounded monthly?

$$t = \frac{\log 2}{12 \log\left(1 + \dfrac{0.112}{12}\right)}$$
$$= 6.22 \text{ years}$$

Figuring Present Value of Single Payment

Because money earns interest, receiving a certain sum in the future is the equivalent of receiving a smaller sum today. Present value formulas here are based on expected earnings only and do not include allowances for such items as the risk of not receiving payment in the future. Other factors, such as decrease in the value of money due to inflation, are assumed to be reflected in the interest rate.

Calculating present value is the reverse of calculating future amount. The latter asks, "If you invest \$100 today at ten percent, what will it be worth in one year?" The answer is \$110. A present value calculation asks, "If your money earns ten percent and you want to receive $110 in one year, how much should you invest today?" The answer is $100. That is, if money earns at the rate of ten percent, then the present value of $110 one year from now is $100.

$$v_p = \frac{p}{\left(1 + \dfrac{r}{c}\right)^{ct}}$$

Example 7.16

A debt of $17,450 is due to be paid to you in fifteen months. The debtor company is reorganizing its liabilities and offers to pay $15,000 now if you will accept it in full payment. If you can invest the money at thirteen percent compounded monthly, should you accept?

To answer, calculate the present value of $17,450, letting $c = 12$, $t = 15/12$ years, and $r = 0.13$.

$$v_p = \frac{17{,}450}{\left(1 + \dfrac{0.13}{12}\right)^{12(15/12)}}$$

$$= \$14{,}845.75$$

You should accept. Another way to answer the question is to use a previous formula to calculate how much $15,000 will grow to by the time the $17,450 is due.

$$a = 15{,}000\left(1 + \frac{0.13}{12}\right)^{12(15/12)}$$

$$= \$17{,}631.31$$

Example 7.17

One of the major applications of the concept of present value is to provide a common reference amount when examining alternatives with different variables.

Suppose you have a contract to deliver a prototype of a large item, and you plan to purchase a subassembly for it. One contractor's assembly will earn $47,834 for you after ten months and another will earn $51,375 after fourteen months. Compare the two purchases.

As both the times and the amounts are different, one way to make the comparison is to calculate the present value of both. Assume that the current interest rate is eleven percent and, to keep the calculation simple (for a comparison), let it be compounded once a year.

$$v_{p1} = \frac{47,834}{(1 + 0.11)^{(10/12)}}$$

$$= \$43,847.96$$

$$v_{p2} = \frac{51,375}{(1 + 0.11)^{(14/12)}}$$

$$= \$45,485.72$$

On this criterion the second contractor's assembly looks better.

PAYMENTS AT REGULAR INTERVALS

The next group of formulas is based on payments, in addition to interest, being added to the fund at regular intervals.

How to Figure Future Payments

When a fixed payment is made each year to a fund that starts at zero, the value of the fund after t years is given by

$$a = s\left[\frac{\left(1 + \dfrac{r}{c}\right)^{ct} - 1}{\dfrac{r}{c}}\right]$$

If interest is compounded annually, $c = 1$ and the formula simplifies to

$$a = s\left[\frac{(1 + r)^t - 1}{r}\right]$$

Example 7.18

A fund for a new lighting system is accumulated by putting $18,000 each year into an investment that pays 9.8 percent compounded annually. How much will be in the fund after five years?

$$a = 18,000\left[\frac{(1 + 0.098)^5 - 1}{0.098}\right]$$

$$= \$109,455.09$$

How Long It Takes a Fund to Reach a Certain Amount

Rearranging the preceding equation yields a formula convenient for calculating how long it takes for the fund to build to a given amount. The full formula is

$$t = \frac{\log(a\,r + s\,c) - \log(s\,c)}{c\,[\log(c + r) - \log c]}$$

Because compounding more than once a year complicates the mathematics considerably, it is frequently accepted that using annual compounding gives results that are close enough.

Annual compounding

$$t = \frac{\log(a\,r + s) - \log s}{\log(1 + r)}$$

The logarithms can have any base as long as the same base is used in each term. Tables often give common logarithms (base 10); calculators are often arranged to make natural logarithms (base e) more convenient.

Example 7.19

The production department wants to outfit a model shop with $32,000 worth of machines as soon as they can. Management will allocate $6,000 per year. If the department accumulates the money in a fund that pays 10.8 percent, how long before they have the $32,000?

$$t = \frac{\log[32,000(0.108) - 6,000] - \log 6,000}{\log(1 + 0.108)}$$

$$= 4.44 \text{ years}$$

The alternative is to purchase $6,000 worth of machines each year, which would take five and a third years. The longer wait for the full complement of machines must be weighed against the fact that some of the machines (presumably those needed the most) would be available sooner.

Present Value of Stream of Payments

If periodic payments are made to a fund, each payment starts earning interest as soon as it is made. In addition, if interest is compounded, all past earnings add to the principal for calculating interest earnings. The next formula calculates the value today of a fund t years in the future.

Full formula

$$v_p = s \left[\frac{1 - \left(1 + \dfrac{r}{c}\right)^{-ct}}{\dfrac{r}{c}} \right]$$

The mathematical complications resulting from compounding are often not justified, as future interest rates are only an estimate. Therefore, it is often accurate enough to assume annual compounding and use the following formula

Simplified formula

$$v_p = s \left[\frac{1 - (1 + r)^{-t}}{r} \right]$$

Example 7.20

When the production department sells part of its fleet of trucks, the buyer offers to pay either $175,000 cash or $45,000 per year for five years. Which offer is better for the seller if you estimate that money will earn 11.5 percent during the five years?

You cannot compare these two forms of payment unless they are expressed in a common value. As the cash offer is already a present value, you should find the present value of the series of payments.

$$v_p = 45,000 \left[\frac{1 - (1 + 0.115)^{-5}}{0.115} \right]$$
$$= \$164,244.50$$

The cash offer, whose present value is $175,000, is better than the series of annual payments whose present value is $164,244.50.

How to Figure Your Regular Annual Payment—With Interest

This formula finds the size of the annual payments that will reduce a debt to zero while including interest on the unpaid balance in each payment.

$$s = p \left[\frac{r}{1 - (1 + r)^{-t}} \right]$$

Example 7.21

The buyers in the preceding example want to determine annual payments that will make the series of payments and the cash payment equally attractive.

They will pay 11.5 percent on the unpaid balance. What should their annual payments be?

$$s = 175{,}000 \left[\frac{0.115}{1 - (1 + 0.115)^{-5}} \right]$$

$$= \$47{,}946.81$$

Calculating Regular Monthly Payments: Two Variables

When the payments are made monthly instead of annually, the same formula is used but there are two differences in the meanings of the variables. First, interest rate r will now be a monthly rate; annual rate divided by twelve. Second, the exponent t will now be the total number of payments instead of the number of years.

Example 7.22

A good customer buys $80,000 of our products. The terms are 10.8 percent per annum on the unpaid balance, monthly payments of principal and interest for four years. What should they pay each month?

$$s = 80{,}000 \left[\frac{\dfrac{0.108}{12}}{1 - \left(1 + \dfrac{0.108}{12}\right)^{-4(12)}} \right]$$

$$= \$2{,}059.88$$

Home mortgages

This formula can also be used to calculate home mortgage payments.

Example 7.23

You have a mortgage of $87,000 and a 13.5 percent loan. What will the monthly payments be toward interest and amortization for a thirty-year loan?

$$s = 87{,}000 \left[\frac{\dfrac{0.135}{12}}{1 - \left(1 + \dfrac{0.135}{12}\right)^{-30(12)}} \right]$$

$$= \$996.51$$

The last computer program in this chapter prints a repayment schedule for monthly payments. It shows the amounts of each payment that cover interest and amortization, and it shows the remaining balance. An example is given, including a print of the table.

Figure 7-1. Computer program solves all formulas in Chapter 7.

```
10 '*****************************************************************
20 '** INTCH7.BAS  Calculates interest equations for Chapter 7  **
30 '** of Handbook of Manufacturing and Production Management   **
40 '** Formulas, Charts, and Tables                             **
50 '*****************************************************************
60 CLEAR,,,32768!: SCREEN 6: COLOR 3,2
70 GOSUB 140            'Display menu
80 CLS
90 ON SELECT GOSUB 330,440,550,660,770,880,990,1100,1210,1310,1430,1550,1670,178
0,1900,2020,2130,2250,2370
100 CLS
110 GOTO 70
120 END
130 'Subroutine to display menu  $$$$$$$$$$$$$$$$$$
140 CLS
150 PRINT
160 PRINT "SIMPLE INTEREST";TAB(40)"COMPOUND INTEREST"
170 PRINT " 1. Earnings";TAB(42)"10. Future value, 1 payment"
180 PRINT " 2. Initial amount";TAB(42)"11. Time required"
190 PRINT " 3. Interest rate";TAB(42)"12. Interest rate"
200 PRINT " 4. Time required";TAB(42)"13. Time for given multiple"
210 PRINT " 5. Total amount";TAB(42)"14. Present value, 1 payment"
220 PRINT " 6. Principal required";TAB(42)"15. Future value, stream"
230 PRINT " 7. Interest rate for total";TAB(42)"16. Time to accumulate, stream"
240 PRINT " 8. Time for total";TAB(42)"17. Present value, stream"
250 PRINT " 9. Time for given multiple";TAB(42)"18. Regular annual payment"
260 PRINT TAB(42)"19. Regular monthly payment"
270 PRINT
280 INPUT "Select by number ",SELECT
290 IF SELECT>0 AND SELECT<20 THEN RETURN
300 PRINT "Select between 1 and 19 only"
310 GOTO 270
320 'Subroutine for earnings, simple interest  11111111111111111
330 PRINT TAB(12) "Calculating earnings, simple interest"
340 GOSUB 2470
350 GOSUB 2490
360 GOSUB 2520
370 E=P*R*T
380 PRINT
390 PRINT "Interest earnings are $";E
400 GOSUB 2900
410 IF TYPE$="A" THEN PRINT: GOTO 330
420 RETURN
430 'Subroutine for initial amount, simple interest  22222222222222
440 PRINT TAB(12) "Calculating initial amount required, simple interest"
450 GOSUB 2610
460 GOSUB 2490
470 GOSUB 2520
480 P=E/(R*T)
490 PRINT
500 PRINT "Initial principal should be $";P
510 GOSUB 2900
520 IF TYPE$="A" THEN PRINT : GOTO 440
530 RETURN
540 'Subroutine for rate of simple interest  3333333333333333
550 PRINT "Calculating rate of simple interest"
560 GOSUB 2610
570 GOSUB 2470
580 GOSUB 2520
590 R=E/(P*T)
```

Figure 7-1. (continued)

```
600 PRINT
610 PRINT "Interest rate will be";R*100;"percent"
620 GOSUB 2900
630 IF TYPE$="A" THEN PRINT : GOTO 550
640 RETURN
650 'Subroutine to calculate time required, simple interest   444444444444
660 PRINT "Calculating time required, simple interest"
670 GOSUB 2610
680 GOSUB 2470
690 GOSUB 2490
700 T=E/(P*R)
710 PRINT
720 PRINT "Investment should continue for";T;"years"
730 GOSUB 2900
740 IF TYPE$="A" THEN PRINT : GOTO 660
750 RETURN
760 'Subroutine to calculate total amount, simple interest   555555555555
770 PRINT "Calculating total amount, simple interest"
780 GOSUB 2470
790 GOSUB 2490
800 GOSUB 2520
810 A=P*(1+R*T)
820 PRINT
830 PRINT "Total amount, including interest earned, is";A
840 GOSUB 2900
850 IF TYPE$="A" THEN PRINT : GOTO 770
860 RETURN
870 'Subroutine to calculate principal required, simple interest   666666666
880 PRINT "Calculating principal required, simple interest"
890 GOSUB 2630
900 GOSUB 2490
910 GOSUB 2520
920 P=A/(1+R*T)
930 PRINT
940 PRINT "Initial principal must be $";P
950 GOSUB 2900
960 IF TYPE$="A" THEN PRINT : GOTO 880
970 RETURN
980 'Subroutine to calculate interest for total amount   777777777777777
990 PRINT "Calculating interest required for total amount, simple interest"
1000 GOSUB 2650
1010 GOSUB 2470
1020 GOSUB 2520
1030 R=(A-P)/(P*T)
1040 PRINT
1050 PRINT "Interest must be";R*100;"percent"
1060 GOSUB 2900
1070 IF TYPE$="A" THEN PRINT : GOTO 990
1080 RETURN
1090 'Subroutine to calculate time to reach given total   88888888888
1100 PRINT "Calculating time to reach given total, simple interest"
1110 GOSUB 2650
1120 GOSUB 2470
1130 GOSUB 2490
1140 T=(A-P)/(P*R)
1150 PRINT
1160 PRINT "Investment must continue for";T;"years"
1170 GOSUB 2900
1180 IF TYPE$="A" THEN PRINT : GOTO 1100
1190 RETURN
```

Figure 7-1. (continued)

```
1200 'Subroutine for time to multiply by given factor  999999999999999
1210 PRINT "Calculating time to multiply by given factor"
1220 GOSUB 2710
1230 GOSUB 2490
1240 T=(M-1)/R
1250 PRINT
1260 PRINT "It will take";T;"years to multiply the principal by";M
1270 GOSUB 2900
1280 IF TYPE$="A" THEN PRINT : GOTO 1210
1290 RETURN
1300 'Subroutine for future value, 1 payment  10 10 10 10 10 10 10
1310 PRINT "Calculating future value, single payment, compound interest"
1320 GOSUB 2670
1330 GOSUB 2490
1340 GOSUB 2830
1350 GOSUB 2520
1360 V=A*(1+R/C)^(T*C)
1370 PRINT
1380 PRINT "Total value of fund will be $";V
1390 GOSUB 2900
1400 IF TYPE$="A" THEN PRINT : GOTO 1310
1410 RETURN
1420 SUBROUTINE FOR TIME REQUIRED, COMPOUND INTEREST  11 11 11 11 11 11 11
1430 PRINT "Calculating time required, compound interest"
1440 GOSUB 2870
1450 GOSUB 2670
1460 GOSUB 2830
1470 GOSUB 2490
1480 T=LOG(V/A)/(C*LOG(1+R/C))
1490 PRINT
1500 PRINT "It will take";T;"years"
1510 GOSUB 2900
1520 IF TYPE$="A" THEN PRINT : GOTO 1430
1530 RETURN
1540 'Subroutine for interest rate, compound interest  12 12 12 12 12 12
1550 PRINT "Calculating interest rate, compound interest"
1560 GOSUB 2830
1570 GOSUB 2870
1580 GOSUB 2670
1590 GOSUB 2520
1600 R=C*((V/A)^(1/(C*T))-1)
1610 PRINT
1620 PRINT "The interest rate must be";R*100;"percent"
1630 GOSUB 2900
1640 IF TYPE$="A" THEN PRINT : GOTO 1550
1650 RETURN
1660 'Subroutine for time to multiply by given factor  13 13 13 13 13 13
1670 PRINT "Calculating time to multiply by given factor, compound interest"
1680 GOSUB 2710
1690 GOSUB 2830
1700 GOSUB 2490
1710 T=LOG(M)/(C*LOG(1+R/C))
1720 PRINT
1730 PRINT "It will take";T;"years to multiply the money by";M
1740 GOSUB 2900
1750 IF TYPE$="A" THEN PRINT : GOTO 1670
1760 RETURN
1770 'Subroutine for present value, single payment  14 14 14 14 14 14
1780 PRINT "Calculating present value, 1 payment, compound interest"
1790 GOSUB 2630
```

Figure 7-1. (continued)

```
1800 GOSUB 2490
1810 GOSUB 2830
1820 GOSUB 2520
1830 V=A/((1+R/C)^(C*T))
1840 PRINT
1850 PRINT "Present value is $";V
1860 GOSUB 2900
1870 IF TYPE$="A" THEN PRINT : GOTO 1780
1880 RETURN
1890 'Subroutine for future value of stream of payments 15 15 15 15 15 15
1900 'PRINT "Calculating future value of stream of payments"
1910 GOSUB 2850
1920 GOSUB 2490
1930 GOSUB 2830
1940 GOSUB 2520
1950 V=S*(((1+R/C)^(C*T)-1)/(R/C))
1960 PRINT
1970 PRINT "Stream of payments will build to $";V
1980 GOSUB 2900
1990 IF TYPE$="A" THEN PRINT : GOTO 1900
2000 RETURN
2010 'Subroutine for time to reach given value  16 16 16 16 16 16 16 16 16
2020 PRINT "Calculating time to accumulate stream of payments"
2030 GOSUB 2870
2040 GOSUB 2490
2050 GOSUB 2690
2060 GOSUB 2830
2070 T=(LOG(V*R+A*C)-LOG(A*C))/(C*(LOG(C+R)-LOG(C)))
2080 PRINT
2090 PRINT "This accumulation will take";T;"years"
2100 GOSUB 2900
2110 IF TYPE$="A" THEN PRINT : GOTO 2020
2120 RETURN
2130 'Subroutine for present value of stream of payments  17 17 17 17 17 17
2140 PRINT "Calculating present value of stream of payments"
2150 GOSUB 2690
2160 GOSUB 2490
2170 GOSUB 2830
2180 GOSUB 2520
2190 V=A*((1-(1+R/C)^(-C*T))/(R/C))
2200 PRINT
2210 PRINT "Present value of stream of payments is $";V
2220 GOSUB 2900
2230 IF TYPE$="A" THEN PRINT : GOTO 2130
2240 RETURN
2250 'Subroutine for regular annual payment  18 18 18 18 18 18 18 18
2260 PRINT "Calculating size of regular annual payment"
2270 GOSUB 2630
2280 GOSUB 2490
2290 GOSUB 2520
2300 S=A*(R/(1-(1+R)^(-T)))
2310 PRINT
2320 PRINT "Annual payment should be $";S
2330 GOSUB 2900
2340 IF TYPE$="A" THEN PRINT : GOTO 2250
2350 RETURN
2360 'Subroutine for regular monthly payment  19 19 19 19 19 19 19 19
2370 PRINT "Calculating regular monthly payment"
2380 GOSUB 2630
2390 GOSUB 2490
```

Figure 7-1. (continued)

```
2400 GOSUB 2520
2410 S=A*((R/12)/(1-(1+R/12)^(-12*T)))
2420 PRINT
2430 PRINT "Monthly payment should be $";S
2440 GOSUB 2900
2450 IF TYPE$="A" THEN PRINT : GOTO 2370
2460 RETURN
2470     INPUT "Enter principal.  Do not use $ or commas. ",P
2480        RETURN
2490 INPUT "Enter interest as a percentage ",R
2500 R=R/100
2510 RETURN
2520     INPUT "Will time be in months or years (M/Y)? ",TQ$
2530     IF TQ$="M" THEN WORD$=" months? ": GOTO 2570
2540     IF TQ$="Y" THEN WORD$=" years? ": GOTO 2570
2550     PRINT "Answer M or Y only"
2560     GOTO 2520
2570     PRINT "How many";WORD$;
2580     INPUT " ",T
2590     IF TQ$="M" THEN T=T/12
2600        RETURN
2610 INPUT "Enter earnings -- do not use $ or commas ",E
2620 RETURN
2630     INPUT "Enter total amount -- do not use $ or commas ",A
2640        RETURN
2650 INPUT "Enter final amount -- do not use $ or commas ",A
2660 RETURN
2670     INPUT "Enter initial amount -- do not use $ or commas ",A
2680        RETURN
2690 INPUT "Enter amount of regular payment -- do not use $ or commas ",A
2700 RETURN
2710 INPUT "By what factor should principal be multiplied? ",M
2720 IF M>0 THEN 2750
2730 PRINT "Multiplier must be greater than 1"
2740 GOTO 2710
2750 PRINT "The original principal will be multiplied by";M
2760 PRINT "Type C for correct, continue with calculation"
2770 INPUT "   or I for incorrect, select another multiplier ",TYPE$
2780 IF TYPE$="C" THEN RETURN
2790 IF TYPE$="I" THEN 2710
2800 PRINT "Type C or I only"
2810 GOTO 2760
2820 RETURN
2830     INPUT "Enter number of compounding periods per year ",C
2840        RETURN
2850 INPUT "Enter amount of regular payment ",S
2860 RETURN
2870     INPUT "Enter total value of fund including compound interest ",V
2880        RETURN
2890 'Subroutine to ask for next step after calculation  $$$$$$$$$$$$$$$
2900 PRINT
2910 PRINT "Type A for another of the same type calculation"
2920 INPUT "  or R to return to menu ",TYPE$
2930 IF TYPE$="A" OR TYPE$="R" THEN RETURN
2940 PRINT "Only A or R should be typed"
2950 GOTO 2900
```

COMPUTER PROGRAMS

Two programs are given; one solves all the formulas in this chapter and the other prints a repayment schedule for the regular monthly payment against principal and interest.

Overall Program for Chapter Formulas

The program listed in Figure 7-1 solves all the formulas in this chapter. It has been tested on an IBM PC, but standard BASIC has been used so it can be used with little or no modification on other computers.

Figure 7-2.   Computer program that prints repayment schedule.

```
10 '***********************************************************
20 '** PAYSKED.BAS   Prints mortgage payback schedule.     **
30 '** For Handbook of Manufacturing and Production Management **
40 '** Formulas, Charts, and Tables                        **
50 '***********************************************************
60 CLEAR,,,32768!: SCREEN 6: COLOR 3,3
70 U$="######.##": UI$="####": UR$="##.##"
80 CLS
90 PRINT "Enter inital amount of debt."
100 INPUT "Do not use $ or commas. ",D
110 INPUT "Enter payback period in months ",T
120 INPUT "Enter interest rate as a percent ",R
130 R=R/1200
140 A=D*(R/(1-(1+R)^(-T)))          'Calculated monthly payment
150 LL=T: ALL=0: FIRST=1: RB=D
160 IF LL>20 THEN 200
170 LT=LL
180 ALL=1
190 GOTO 220
200 LT=20
210 LL=LL-20
220 CLS
230 PRINT "Original debt = $";
240 PRINT USING U$;D;
250 PRINT TAB(35)"Monthly payment = $";
260 PRINT USING U$;A
270 PRINT "Interest rate =";
280 PRINT USING UR$;R*1200;
290 PRINT "%"
300 PRINT "Number";TAB(15)"Interest";TAB(30)"Amortization";TAB(45)"Balance"
310 FOR C=FIRST TO FIRST+LT-1
320    I=RB*R
330    AM=A-I
340    IF AM>RB THEN AM=RB
350    PRINT USING UI$;C;
360    PRINT TAB(15) USING U$;I;
370    PRINT TAB(30) USING U$;AM;
380    RB=RB-AM
390    PRINT TAB(45) USING U$;RB
400 NEXT C
410 INPUT "Press <Enter> to continue",E$
420 IF ALL=1 THEN 450
430 FIRST=FIRST+20
440 GOTO 160
450 PRINT T;"payments have been shown."
460 INPUT "Press <Enter> for another table.",E$
470 GOTO 80
```

The program is structured with a subroutine handling each choice in the menu. Each subroutine is identified with a series of numbers that are the same as the number of the choice. Therefore if any part of the program malfunctions, it will be easy to narrow the problem down to a small group of lines. Proofread them carefully against the listing in the book.

The program also includes checks against operator error when running. For example, there are nineteen items to choose from in the menu; if someone accidentally chooses number 29, line 290 will reject the choice, line 300 will explain the error, and line 310 will send the program back to give the operator another chance.

The menu items are arranged in the same order as their respective formulas in the chapter. Therefore it is easy to check the program after entering it into your computer; go through the chapter and try each example on the computer. If any answer differs from the book's answer, check the subroutine that handles that formula.

Monthly Payments Program

Figure 7-2 lists the program that prints a table showing the status of the debt after each payment. It first calculates the monthly payment and then it prints the table showing:

Table 7-1. Example of table printed by program in Figure 7-2.

Original debt = $21000.00
Interest rate = 14.10%

Monthly payment = $1301.21

Number	Interest	Amortization	Balance
1	246.75	1054.46	19945.54
2	234.36	1066.85	18878.70
3	221.82	1079.38	17799.32
4	209.14	1092.06	16707.25
5	196.31	1104.90	15602.36
6	183.33	1117.88	14484.48
7	170.19	1131.01	13353.46
8	156.90	1144.30	12209.16
9	143.46	1157.75	11051.41
10	129.85	1171.35	9880.06
11	116.09	1185.12	8694.94
12	102.17	1199.04	7495.90
13	88.08	1213.13	6282.77
14	73.82	1227.38	5055.39
15	59.40	1241.81	3813.58
16	44.81	1256.40	2557.18
17	30.05	1271.16	1286.02
18	15.11	1286.02	0.00

Press <Enter> to continue

the number of each payment

the part of each payment that covers interest

the part of each payment that reduces the principal

the principal remaining after each payment

Example 7.24

A debt of $21,000 is to be paid in eighteen monthly installments, along with 14.1 percent on the unpaid balance.

The program first asks for these amounts. Then it calculates that the monthly payment should be $1,301.21. It then prints Table 7-1.

Rounding may cause the remaining balance to drift slightly during the life of the loan. The program adjusts if a small negative balance would remain after the last payment. Line 340 checks and if the amount marked for reducing the principal is more than the last principal, that line changes the last payment to make the balance end with zero. In the example, the last payment (sum of the middle two columns) will be $1,301.13.

CHAPTER 8

Calculations Related to Profit, Rent or Buy, Investment Decisions, and Depreciation

The production department is a source of operating profit, and is responsible for most decisions related to capital equipment. This chapter looks at some of the key ways to evaluate profit and change in profit, as well as investment and depreciation calculations. Should you rent or buy? Which is more cost effective? Formulas that answer these and more are included here.

You'll learn the basic components of profit and how they affect production, how to analyze the present value of a rental, what the "indifference point" is and how to calculate it, the breakeven service life, and three key investment decisions that are widely used when viewing important investment decisions.

Symbols used

b = annual benefit from use of capital equipment

c_f = fixed costs

c_t = present value of total costs

c_v = variable costs—unit variable costs times number produced

d = depreciation

d_p = decimal or percentage change in profit

f_r = annual fee for equipment rental

f_{rb} = breakeven rental fee

l = length of time you have a piece of capital equipment

m = annual maintenance and operating costs

n = number of units sold

n_b = breakeven point in number of units sold

p = price of capital equipment

p_d = profit per dollar of sales

p_o = operating profit, due to sales of product

p_0 = profit due to sales in year 0

p_1 = profit due to sales in year 1

r = interest rate, decimal

r_p = present value of rental arrangement

r_s = revenue from sales—unit price times number sold

r_t = amount of capital investment remaining to be recouped after t years

s = service life of equipment

s_b = breakeven, or indifference, service life

t = time—number of years

t_p = time to pay back capital investment

u_f = unit contribution to fixed costs

PROFIT: THE BASIC COMPONENTS

Business managers, economists, politicians, and voters all have different concepts and definitions of profit. Let's look at a few components of profit.

Operating Profit: Fixed and Variable Costs

In broad terms, operating profit is the money left from sales revenue after subtracting fixed costs and variable costs. Fixed costs are those that do not change with the level of production, such as top executives' salaries and interest on outstanding debt. Variable costs, such as the cost of material, depend on the amount produced.

$$p_o = r_s - c_f - c_v$$

Example 8.1

A toy manufacturer sells 65,000 dolls at $12.75 each. Material, direct labor, and other variable costs are $5.40 for each doll. Property taxes, interest on debt, and other fixed costs are $410,192. What is their operating profit?

$$p_o = 12.75(65,000) - 410,192 - 5.40(65,000)$$
$$= \$67,558$$

Unit Profit

In addition to total operating profit, financial analysis can include profit per unit. It is simply total operating profit divided by the number of units.

$$p_{01} = \frac{\text{operating profit}}{\text{number of units}}$$

If the number of units sold is different from the number of units produced, then either one can be used in the formula, depending on the type of analysis being made.

Example 8.2

The toy manufacturer in the preceding example wants to know how much each doll sold contributes to profit.

$$p_{01} = \frac{67,558}{65,000}$$
$$= \$1.04$$

Covering Fixed Costs

Minimum selling price

Whereas *total* fixed costs is a constant, the amount of fixed costs covered by *each* sale depends on the number of sales. An important prerequisite to reducing prices is to reduce the portion of total fixed costs that each unit sold must carry. It sets a minimum selling price for keeping the business going without a loss. Even if there were no variable costs and no profit, the business would eventually fail if it did not cover its fixed costs.

$$u_f = \frac{r_s - c_v - p_o}{n}$$

Sales of product is emphasized because nonoperating sales, such as selling a piece of real estate that had been intended for a new plant, should not enter into this analysis.

Example 8.3

On total sales of $1,237,482 from 411,563 units a firm has variable costs of $517,406. It expects to make a profit of $190,000. What is each unit's contribution to fixed costs?

$$u_f = \frac{1,237,482 - 517,406 - 190,000}{411,563}$$

$$= \$1.29$$

Profit per Dollar of Sales

Another figure of interest is found by dividing total sales in dollars by total profit.

$$p_d = \frac{\text{total profit}}{\text{total sales in dollars}}$$

Example 8.4

What is the profit per dollar of sales for the firm in the preceding example?

$$p_d = \frac{190,000}{1,237,482}$$

$$= \$0.15$$

Percentage Change in Profit

When sales change by a certain percentage, the presence of fixed costs means profit will not change by the same percentage.

$$d_p = \frac{p_0 - p_1}{p_0}$$

If change in profit is to be expressed as a percent, then d_p should be multiplied by 100. Notice that this figure will be positive if profit increases, and negative if it decreases.

Example 8.5

Sales of the toy manufacturer in the first example drop ten percent, to 65,000 minus 6,500 or 58,500. If the unit price did not change, what is the percentage change in their profit?

Profit in the second year is

$$p_1 = 12.75(58,500) - 410,192 - 5.40(58,500)$$

$$= \$19,783$$

The percentage change in profit has been

$$d_p = \frac{19,783 - 67,558}{67,558}$$

$$= -71 \text{ percent}$$

The large amount of fixed costs has caused a ten percent drop in sales to result in a seventy-one percent drop in profit.

Determining the Breakeven Point

The sales volume that results in zero loss and zero profit is found from the following formula. Selling less than this amount results in a loss; profits increase with sales beyond the breakeven point.

$$n_b = \frac{\text{total fixed costs}}{\text{unit selling price} - \text{variable costs per unit}}$$

Example 8.6

Fixed costs are \$310,000 and variable costs are \$73.18 per unit. How many units must be sold for the firm to break even if the units sell for \$125.00 each?

$$n_b = \frac{310,000}{125.00 - 73.18}$$

$$= 5,982 \text{ units}$$

CAPITAL EQUIPMENT—DECIDING WHETHER TO RENT OR BUY

Present value concept

This section approaches this decision from several directions. The first has been developed from the concept of *present value*, covered in Chapter 7.

It is possible for the present value of a capital investment to have a negative value, indicating that it would be better to use the funds in another way. However, as long as a rented investment earns more than its rental fee, it has a positive value because it does not lock in funds that have alternative uses.

Analyzing the Present Value of a Rental

Money paid to rent equipment is paid currently and does not have to be adjusted to present value. However, the overall rental arrangement can be the equivalent of a certain amount of money today.

$$r_p = \frac{b - f_r}{r}$$

Annual benefit, b, is the dollar improvement in profit, cash flow, or other measures that result from renting the equipment. The value of the rent or buy analysis depends on how realistically and accurately this figure is estimated.

The last variable, r, can be an interest rate—the rate that would be paid if the funds were kept in an interest-bearing investment. It can also be the earning rate of alternative capital equipment.

Example 8.7

Find the present value of a machine that rents for $3,150 annually if it adds $3,970 to profit while interest rates are 10.5 percent. As with other calculations, the rate given in percent is first converted to a decimal by dividing it by 100.

$$r_p = \frac{3,970 - 3,150}{0.105}$$

$$= \$7,809.52$$

One of the main reasons for finding the present value equivalent is to bring alternatives with different variables to a common dimension. Alternatives frequently cover different lengths of time, return different benefits, and cost different amounts. They can be compared by expressing them all in present value.

Breakeven Rental Fee

Use the present value of a capital investment, from Chapter 7, and the present value of a rental, from the beginning of this chapter, to determine the breakeven rental fee. When preparing for negotiating meetings, it would be helpful to determine this indifference point—the rental fee that makes the benefit of renting equal to the benefit of buying. Any smaller rental fee makes renting more attractive; any larger rental fee makes buying more attractive.

The indifference point

$$f_{rb} = \frac{b}{(1 + r)^s} + r\,p$$

Total price of the equipment should include costs such as shipping, maintenance, property taxes, and so forth. Costs that apply to both renting and buying, such as reinforcing the factory floor, training operators cancel and can be left out of the analysis.

Example 8.8

A machine with a six-year service life can be purchased for $82,000 or rented. If it will contribute $9,000 a year to profit, and the current interest rate is 9.7 percent, what is the maximum annual rental fee we should consider?

$$f_{rb} = \frac{9,000}{(1 + 0.097)^6} + 0.097(82,000)$$

$$= \$13,118.20$$

That rental fee leaves us indifferent as to whether renting or buying places us in a better position.

Breakeven Service Life

In many situations an equation that gives the breakeven service life can be more meaningful to the production executive. Because s is an exponent, solving for it involves logarithms; this equation has become more popular as there is a computer or scientific calculator on every desk. This calculation is included in the computer program at the end of this section.

$$s_b = \frac{\log b - \log(f_r - rp)}{\log(1 + r)}$$

This is our indifference service life; any service life longer than s_b makes buying more attractive. Logarithms to any base can be used as long as the same base is used throughout the calculation. Many calculators place natural logarithms on a main key and give common logarithms a second function position, making the former more convenient. The BASIC language makes natural logarithms more convenient in computer programs.

Two Traps to Avoid

Trap 1: anti-logarithms

You should avoid two traps this equation presents. First, notice that the ratio of logarithms is the answer; do not take the anti-logarithm. This is an easy mistake to make because most logarithm problems are not finished until you take the anti-logarithm.

Trap 2: partial results

The second trap exists because the numerator can sometimes be a small difference between relatively large numbers. Therefore, if you write down partial results and then reenter them in your calculator, be sure not to round the partial results. There are four ways you can avoid this type of error.

1. Perform the entire calculation as a continuous, chained operation on the calculator.

2. If you are less familiar with using all the features of your calculator, and you prefer to write down partial results, such as the numerator, write down all the decimal places your calculator gives. Then, when reentering the partial result, include all the decimal places.

3. If your calculator has several memories, store partial results and then recall them for the final calculation. Calculators store quantities in memory with all their decimal places.

4. Use the computer program listed in the next section. It completes the calculations without rounding.

Example 8.9

A machine that will increase profits by \$11,000 a year rents for \$53,000. If you buy the machine, its total cost will be \$420,000. Find the breakeven service life during a time when the applicable interest rate is 10.8 percent.

$$s_b = \frac{\log 11{,}000 - \log[53{,}000 - .01(10.8)420{,}000)]}{\log[1 + .01(10.8)]}$$

$$= 3.55 \text{ years}$$

COMPUTER PROGRAM FOR RENT OR BUY CALCULATIONS

All the formulas in this rent or buy section are included in the program listed in Figure 8-1. After each calculation it returns to the menu and allows you to select which of the formulas to calculate next.

After you enter the program in your computer, test it with the example given for each of the formulas. The program is structured with a subroutine assigned to each formula, so if any answer is wrong, you can easily determine the program area where the problem is. Carefully proofread all lines in that area against the listing in Figure 8-1.

THREE KEY INVESTMENT DECISIONS

From the several known methods of viewing investment decisions, three have been selected that are widely used.

1. Payback period, short term

It is expected that the purchase of capital of equipment will result in the firm earning more than it would without the equipment. This formula calculates how long it takes for the purchase price to be returned through extra earnings.

Figure 8-1. Computer program for rent or buy calculations.

```
10 '*******************************************
20 '** RENTBUY.BAS  Rent or buy calculations for   **
30 '** Handbook of Manufacturing and Production    **
40 '** Management Formulas, Charts, and Tables      **
50 '*******************************************
60 CLEAR,,,32768!: SCREEN 6: COLOR 3,3
70 CLS
80 GOSUB 1010          'Display menu
90 CLS
100 INPUT "Enter annual benefit ",B
110 PRINT
120 PRINT "Enter applicable interest rate in"
130 INPUT "percent -- not converted to a decimal ",I
140 PRINT
150 ON SELECT GOSUB 1210,1410,1410
160 GOTO 70
170 END
1000 'Subroutine to display menu and accept operator choices  <<<<<<<<<<
1010 PRINT
1020 PRINT "1. Present value of rental"
1030 PRINT
1040 PRINT "2. Breakeven rental fee"
1050 PRINT
1060 PRINT "3. Breakeven service life"
1070 PRINT
1080 INPUT "Select by number ",SELECT
1090 IF SELECT>0 AND SELECT<4 THEN RETURN
1100 PRINT "Select between 1 and 4 only"
1110 GOTO 1070
1200 '(1) Subroutine to calculate present value of rental  111111111111
1210 INPUT "Enter annual rental fee ",F
1220 PRINT
1230 RP=(B-F)/(.01*I)
1240 PRINT "When rental fee, annual benefit, and"
1250 PRINT "interest rate are $";F;", $";B;", and";I;"%,"
1260 PRINT "the rental's present value is $";RP
1270 PRINT
1280 INPUT "Press ENTER to return to menu ",E$
1290 RETURN
1400 '(2) Subroutine to calculate breakeven rental fee  222222222222
1410 INPUT "Enter total cost of buying equipment ",P
1420 PRINT
1430 IF SELECT=3 THEN GOSUB 1800 ELSE GOTO 1450
1440 GOTO 1570
1450 INPUT "Enter years of service life ",S
1460 PRINT
1470 FB=B/((1+.01*I)^S)+.01*I*P
1480 PRINT "When purchase cost, service life,"
1490 PRINT "annual benefit, and interest rate"
1500 PRINT "are $";P;",";S;"years,"
1510 PRINT "$";B;", and";I;"%, a fee of"
1520 PRINT "$";FB;"makes renting and buying"
1530 PRINT "equally attractive."
1540 PRINT
1560 INPUT "Press ENTER to return to menu ",E$
1570 RETURN
1800 '(3) Subroutine to calculate breakeven service life  3333333333333
1810 INPUT "Enter annual rental fee ",F
1820 PRINT
1830 PRINT "When rental fee, purchase price, annual"
```

Figure 8-1. (continued)

```
1840 PRINT "benefit, and interest rate are $";F;", $";P;","
1850 PRINT "$";B;", and";I;"%, a service life which"
1860 PRINT "makes renting and buying equally attractive"
1870 IF .01*I*P>=F THEN 1920
1880 LB=(LOG(B)-LOG(F-.01*I*P))/LOG(1+.01*I)
1890 IF LB<0 THEN 1920
1900 PRINT "is";LB;"years."
1910 GOTO 1930
1920 PRINT "does not exist."
1930 PRINT
1940 INPUT "Press ENTER to return to menu ",E$
1950 RETURN
```

To keep the formula simple, for a first analysis, we omit any correction for changes in the value of money. This correction would be small if the payback period is short or if alternative uses of the money would earn very little.

$$t_p = \frac{p - s}{b}$$

The amount paid should include all costs that would occur if the capital investment were not being made. For example, shipping, preparing the factory floor, insurance premiums, and training operators should be included.

Likewise, benefits should include all money inflow that would not occur if the capital investment were not being made. They should include increases in profit, improvements in cash flow, speedier response to customers' requests, and other items. Estimating the dollar value of some of the benefits is the most challenging part of using this formula.

Example 8.10

You are considering the purchase of a machine for $41,000. It will require another $7,000 to bring electric power to the area and to install a branch of the main conveyor belt to carry its output. You expect to sell it in a few years for $13,000. If your annual profits improve by $5,800, how long will it take to recover the cost of this investment?

Purchase of this machine commits you to spend another $7,000, which should be added to the purchase price for this formula.

$$t_p = \frac{41,000 + 7,000 - 13,000}{5,800}$$

$$= 6.03 \text{ years}$$

You can add some refinement to this formula by considering that b represents after-tax net benefits.

2. Payback to present value

If the value of money changes significantly during the payback period, it is worth adding that correction to the formula. The mathematical procedure to follow is to calculate, after each year of service life, how much of the original investment is still to be recouped. This amount is calculated in present value. Then, the year that the remaining amount turns negative is the year in which the original investment is recouped.

$$\begin{matrix} \text{Remaining to recoup} \\ \text{after year } t \end{matrix} = \left(\begin{matrix} \text{Original} \\ \text{invest-} \\ \text{ment} \end{matrix} \right) - \left(\begin{matrix} \text{Present} \\ \text{value of} \\ \text{salvage} \end{matrix} \right) - \left(\begin{matrix} \text{Present value of} \\ \text{annual benefits} \\ \text{through year } t \end{matrix} \right)$$

$$r_t = p - \frac{s}{(1 + r)^l} - b \left[\frac{1 - (1 + r)^{-t}}{r} \right]$$

The computer program at the end of this section will print the remaining value at the end of each year. When the remaining value becomes negative, the program will determine the point in the year when the investment was paid off.

Example 8.11

Total installed and operating price of a proposed machine is $72,000. You expect to sell it for $23,000 in ten years. The machine will yield benefits of $19,500 each year. Assuming that 12.5 percent represents interest rates for the entire life of the machine, what is the payback period?

Instead of starting with the first year, you can avoid a few calculations by first estimating the year in which payback is completed. If the result is positive, try another year. Additional years are easier to calculate because the first two terms are constant and have to be calculated only once. That is, you have to calculate only the third term for each year.

To demonstrate, let's calculate the amount remaining after three years.

$$r_3 = 72,000 - \frac{23,000}{(1 + 0.125)^{10}} - 19,500 \left[\frac{1 - (1 + 0.125)^{-3}}{0.125} \right]$$

$$= 72,000 - 7,082.76 - 46,436.21$$

$$= 64,917.24 - 46,436.21$$

$$= \$18,481.03$$

As this result is positive, you have to continue on to the fourth year. However, the algebraic sum of the first two terms, 64,917.24, does not change—you have to recalculate only the third term and subtract it from that figure.

$$r_4 = 64,917.24 - 19,500 \left[\frac{1 - (1 + 0.125)^{-4}}{0.125} \right]$$

$$= \$6,307.27$$

Payback is not complete, so let's calculate the fifth year.

$$r_5 = 64,917.24 - 19,500 \left[\frac{1 - (1 + 0.125)^{-5}}{0.125} \right]$$

$$= \$ - 4,513.84$$

The negative result tells us that payback was completed in the fifth year. In addition, because the absolute value of r_4 is larger than that of r_5, you can conclude that payback was completed in the second half of year five. The computer program will give the fraction of a year, based on linear interpolation. (It will show that this payback is completed in 4.58 years.)

3. Present value lifetime cost

A realistic way to question the purchase of capital equipment is to determine all costs for the life of the equipment, converted to present value. This analysis is especially useful for comparing potential investments, all of which may have different lifetimes, prices, maintenance requirements, and other variables.

Start with initial cost (given at present value), then add annual operating and maintenance cost (adjusted for present value of a stream of payments), and subtract salvage value (adjusted for present value of a single payment). Present value adjustments are explained in Chapter 7.

$$c_t = p + m \left[\frac{1 - (1 + r)^{-t}}{r} \right] - s (1 + r)^{-t}$$

Total paid should include all amounts incidental to the purchase. If a continuing training program is required, it should be included with m.

Example 8.12

To be ready for any challenges when you argue for a new stamping machine at the next budget meeting, you are going to figure its lifetime cost. Purchase and all preparations for operation will cost $38,750. After that it will cost $4,230 every year to operate and maintain, until it is sold for $4,050 after six years. An earning rate of 12.8 percent applies.

$$c_t = 38,750 + 4,230 \left[\frac{1 - (1 + 0.128)^{-6}}{0.128} \right] - 4,050 (1 + 0.128)^{-6}$$

$$= \$53,788.16$$

If it is not reasonable to use a constant for annual operation and maintenance cost, then calculate the present value of each year's cost. Add those present values, and use their sum in place of the second term of the formula.

If lifetime cost is being used to compare two or more investments with different lifetimes, one method often accepted is to find the arithmetic mean (average) annual cost of each by dividing each c_t by its lifetime.

Investment Decision Computer Program

Figure 8-2 lists a computer program that solves the formulas in this section. All applicable questions are asked on the screen and then the computer prints the answer.

Use the examples given by the formulas to test the program after you enter it. The program is structured with a subroutine for each calculation; if any example does not work right, you can easily find the series of lines for that particular type of calculation.

DEPRECIATION

Because depreciation is an annual allocation of the cost of equipment with a lifetime of more than one year, the method used can significantly affect profits. Tax legislation encourages a variety of depreciation schedules that may or may not continue. However, the methods described here are the traditional ones that are likely to remain through many tax law changes.

Straight-Line Depreciation

Fixed dollar amount

This is the simplest method. Net cost of a piece of capital equipment is divided by the appropriate number of years, and the book value is reduced by the same amount each year.

$$d = \frac{p - s}{t}$$

Example 8.13

Set up a straight-line depreciation schedule for a machine that costs $273,000 and has a salvage value of $25,000. The depreciation period will be eight years.

$$d = \frac{273,000 - 25,000}{8}$$

$$= \$31,000$$

Having determined the amount to depreciate each year, you can prepare a table for the eight years. Each year, subtract $31,000 from the previous year's book value, as shown in Table 8-1. The book value depreciates to the salvage value.

Figure 8-2.   Computer program for investment decision calculations.

```
10 '********************************************************
20 '** INVDEC.BAS  Solves investment decision equations    **
30 '** in Finance chapter of Handbook of Production         **
40 '** and Manufacturing Management Formulas, Charts,       **
50 '** and Tables                                           **
60 '********************************************************
100 CLEAR,,,32768!: SCREEN 6: COLOR 3,6
110 DIM R(100)
120 UY$="###": UR$="#######.##"
200 CLS
210 GOSUB 1010              'Display menu
220 CLS
230 'Enter values used by all formulas  <<<<<<<<<<<<<<
240 PRINT "Enter cost to buy, install, and start using"
250 INPUT "the capital equipment ",P
260 PRINT
270 INPUT "Enter the salvage value ",S
280 PRINT
290 IF P=>S THEN 370
300 PRINT "Purchase cost is less than salvage value."
310 INPUT "Is that what you intend?  (Y/N) ",PS$
320 PRINT
330 IF PS$="Y" THEN 370
340 IF PS$="N" THEN 240
350 PRINT "Answer Y or N please!"
360 GOTO 300
370 ON SELECT GOSUB 2110,2310,2710
380 GOTO 200
1000 'Subroutine to display menu  <<<<<<<<<<<<<<<<,
1010 PRINT
1020 PRINT "1 Simple payback"
1030 PRINT
1040 PRINT "2 Payback to present value"
1050 PRINT
1060 PRINT "3 Present value lifetime cost"
1070 PRINT
1080 INPUT "Select by number ",SELECT
1090 IF SELECT>0 AND SELECT<4 THEN RETURN
1100 PRINT "Select only 1 through 3"
1110 GOTO 1070
1200 'Subroutine to enter benefit amount  <<<<<<<<<<<<<<
1210 INPUT "Enter annual benefit ",B
1220 IF B<P THEN 1290
1230 PRINT "Annual benefit is larger than purchase"
1240 INPUT "price.  Is that what you intended?  (Y/N) ",BQ$
1250 IF BQ$="Y" THEN 1290
1260 IF BQ$="N" THEN 1210
1270 PRINT "Answer Y or N please!"
1280 GOTO 1230
1290 PRINT
1300 RETURN
1400 'Subroutine to enter interest rate  <<<<<<<<<<<<<<<<
1410 INPUT "Enter annual rate of interest as a percentage ",I
1420 IF I<50 THEN 1490
1430 PRINT "That's a very high rate of interest.  Is it"
1440 INPUT "the rate you intended?  (Y/N) ",IQ$
1450 IF IQ$="Y" THEN 1490
1460 IF IQ$="N" THEN 1410
1470 PRINT "Answer Y or N please!"
1480 GOTO 1430
```

Figure 8-2. (continued)

```
1490 PRINT
1500 RETURN
1600 SUBROUTINE TO ENTER TIME  <<<<<<<<<<<<<<<<
1610 INPUT "Enter lifetime of investment in years ",L
1620 IF L>20 THEN 1650
1630 IF L<1 THEN 1710
1640 GOTO 1770
1650 PRINT "That's a long lifetime for most capital equipment."
1660 INPUT "Are you sure it's right?  (Y/N) ",LQ$
1670 IF LQ$="Y" THEN 1770
1680 IF LQ$="N" THEN 1610
1690 PRINT "Answer Y or N please! "
1700 GOTO 1250
1710 PRINT "Capital equipment does not usually have a lifetime"
1720 INPUT "less than one year.  Is your value correct?  (Y/N) ",LQ$
1730 IF LQ$="Y" THEN 1770
1740 IF LQ$="N" THEN 1610
1750 PRINT "Answer Y or N please!"
1760 GOTO 1710
1770 PRINT
1780 RETURN
1900 'Subroutine to enter operation and maintenance cost  <<<<<<<<<<<<
1910 INPUT "Enter annual amount for operation and maintenance ",M
1920 IF M<P THEN 1990
1930 PRINT "You show annual maintenance larger than purchase"
1940 INPUT "price.  Are your figures correct?  (Y/N) ",MQ$
1950 IF MQ$="Y" THEN 1990
1960 IF MQ$="N" THEN 1910
1970 PRINT "Answer Y or N please!"
1980 GOTO 1930
1990 PRINT
2000 RETURN
2100 '(1) Subroutine to calculate short term analysis  11111111111111
2110 GOSUB 1210          'Enter B
2120 TP=(P-S)/B
2130 PRINT "Short term time to recover investment is";TP;"years."
2140 PRINT
2150 INPUT "Press ENTER to return to menu ",E$
2160 RETURN
2300 '(2) Subroutine to calculate payback to present value  222222222222
2310 GOSUB 1210
2320 GOSUB 1410
2330 GOSUB 1610
2340 PRINT
2350 PRINT "End of year       Remaining to recoup"
2360 T=1
2370 T2=S/((1+I*.01)^L)
2380 T3=B*((1-(1+I*.01)^(-T))/(I*.01))
2390 R(T)=P-T2-T3
2400 PRINT TAB(3) USING UY$;T;
2410 PRINT TAB(20) USING UR$;R(T)
2420 IF R(T)>0 THEN 2470
2430 IF R(T)=0 THEN 2450
2440 GOTO 2500
2450 PRINT
2460 GOTO 2400
2470 T=T+1
2480 GOTO 2370
2490 PRINT
2500 RT=T-1+(R(T-1)/(ABS(R(T))+ABS(R(T-1))))
```

Figure 8-2. (continued)

```
2510 PRINT "If interest rate is";I;"percent, the investment"
2520 PRINT "whose price, salvage value, annual benefit, and"
2530 PRINT "service life are $";P;", $";S;", $";B;","
2540 PRINT "and";L;", the payback period is";RT;"years."
2550 INPUT "Press ENTER to return to menu ",E$
2560 RETURN
2700 '(3) Subroutine to calculate present value lifetime cost   33333333333
2710 GOSUB 1410
2720 GOSUB 1610
2730 GOSUB 1910
2740 T2=M*((1-(1+.01*I)^(-L))/(.01*I))
2750 T3=S*(1+.01*I)^(-L)
2760 CT=P+T2-T3
2770 PRINT "When purchase price is $";P;", annual maintenance cost"
2780 PRINT "is $";M;", salvage value is $";S;", interest rate is";I
2790 PRINT "percent, and the equipment's lifetime is";L;"years,"
2800 PRINT "present value lifetime cost is $";CT
2810 PRINT
2820 INPUT "Press ENTER to return to menu ",E$
2830 RETURN
```

Declining Balance Depreciation

Fixed percentage

Instead of a fixed dollar amount each year, this method takes a fixed percentage of the remaining book value. The dollar amount will therefore decline each year. Except for certain restrictions, such as the maximum percentage allowed by tax laws, the fixed percentage is found by doubling the straight-line rate. Salvage value is not considered when using declining balance.

Example 8.14

A production machine bought for $83,000 is expected to have a service life of eleven years. Show its declining balance schedule for those years.

Straight-line depreciation would take off $83,000/11, or $7,545.45 each year, which is 9.09 percent. The declining balance will use double that rate, or 18.18 percent of the remaining balance each year.

Table 8-2 shows the schedule. The first year's depreciation is 18.18 percent

Table 8-1.   Straight-line depreciation.

Year	Depreciation	Book value at end of year
1	31000.00	217,000.00
2	31000.00	186,000.00
3	31000.00	155,000.00
4	31000.00	124,000.00
5	31000.00	93,000.00
6	31000.00	62,000.00
7	31000.00	31,000.00
8	31000.00	0.00

Table 8-2.   Declining balance depreciation.

Annual factor = .1818

Year	Depreciation	Book value at end of year
	Initial value $83,000.00	
1	15089.40	67910.60
2	12346.15	55564.46
3	10101.62	45462.84
4	8265.14	37197.69
5	6762.54	30435.15
6	5533.11	24902.04
7	4527.19	20374.85
8	3704.15	16670.70
9	3030.73	13639.97
10	2479.75	11160.22
11	2028.93	9131.30

of the original amount of $83,000. Then each year's depreciation is 18.18 percent of the balance at the end of the preceding year.

One advantage of this method is that depreciation is faster during early years, making book value closer to actual market value. However, unlike the straight-line method, declining balance never completely writes off the balance.

Sum-of-Years Digits Depreciation

The depreciation factor

With this method, neither the dollar amount nor the percentage remain constant. The depreciation factor is a fraction whose denominator is the sum of the years of service life. Each year's numerator is one of the years, starting with the highest. For example, if the investment has a five year service life, the permanent denominator is 1 + 2 + 3 + 4 + 5, or 15. The first year's numerator is

Table 8-3.   Sum-of-years digits depreciation.

Enter initial cost 22000
Enter length of write-off period 6
Enter salvage value 4000

Year	Fraction	Depreciation	Book value at end of year
		Initial value $18000	
1	6/21	5142.86	12857.14
2	5/21	4285.71	8571.43
3	4/21	3428.57	5142.86
4	3/21	2571.43	2571.43
5	2/21	1714.29	857.14
6	1/21	857.14	0.00

Figure 8-3.   Computer program for depreciation.

```
10 '*************************************************************
20 '** DEPRTN.BAS  Depreciation calculations for Handbook    **
30 '** of Manufacturing and Production Management            **
40 '** Formulas, Charts, and Tables                          **
50 '*************************************************************
60 CLEAR,,,32768!: SCREEN 6: COLOR 3,9
70 UI$="####": UD$="#######.##"
80 CLS
90 GOSUB 1010
100 CLS
110 INPUT "Enter initial cost ",PURCH
120 PRINT
130 INPUT "Enter length of write-off period ",T
140 PRINT
150 ON SELECT GOSUB 1210,1410,1610
160 GOTO 70
170 END
1000 'Subroutine to display menu   <<<<<<<<<<<<<<<<<,
1010 PRINT
1020 PRINT "1 Straight-line depreciation"
1030 PRINT
1040 PRINT "2 Declining balance depreciation"
1050 PRINT
1060 PRINT "3 Sum-of-years digits depreciation"
1070 PRINT
1080 INPUT "Select by number ",SELECT
1090 IF SELECT>0 AND SELECT<4 THEN RETURN
1100 PRINT "Select 1, 2, or 3 only!"
1110 GOTO 1020
1200 '(1) Subroutine to calculate straight-line depreciation  1111111111
1210 INPUT "Enter salvage value ",SALVAGE
1220 CLS
1230 GOSUB 2010
1240 DEC=(PURCH-SALVAGE)/T
1250 PURCH=PURCH-SALVAGE
1260 FOR TC=1 TO T
1270    PURCH=PURCH-DEC
1280    PRINT TAB(2) USING UI$; TC;
1290    PRINT TAB(14) USING UD$; DEC;
1300    PRINT TAB(32) USING UD$; PURCH
1310 NEXT TC
1320 INPUT "Press ENTER to return to menu ",E$
1330 RETURN
1400 '(2) Subroutine to calculate declining balance depreciation  2222222222
1410 FI=2/T
1420 ANNUALFAC=INT(FI*10^4)/10^4
1430 PRINT "Annual factor =";ANNUALFAC
1440 GOSUB 2010
1450 FOR TC=1 TO T
1460    PRINT TAB(2) USING UI$;TC;
1470    PRINT TAB(14) USING UD$;ANNUALFAC*PURCH;
1480    PRINT TAB(32) USING UD$;PURCH-ANNUALFAC*PURCH
1490    PURCH=PURCH-ANNUALFAC*PURCH
1500 NEXT TC
1510 INPUT "Press ENTER to return to menu ",E$
1520 RETURN
1600 '(3) Subroutine to calculate sum-of-years digits depreciation  33333333
1610 DIG=0
1620 INPUT "Enter salvage value ",SALVAGE.
1630 FOR TC=1 TO T
```

Figure 8-3. (continued)

```
1640    DIG=DIG+TC
1650 NEXT TC
1660 PRINT TAB(38)"Book value at"
1670 PRINT "year";TAB(10)"Fraction";TAB(22)"Depreciation";TAB(38)"End of year"
1680 PURCH=PURCH-SALVAGE
1690 PD=PURCH
1700 PRINT TAB(13)"Initial value -- $";PURCH
1710 FOR TC=1 TO T
1720    DECL=((T-TC+1)/DIG)*PURCH
1730    PD=PD-DECL
1740    IF PD<.005 THEN PD=0
1750    PRINT TAB(2) USING UI$;TC;
1760    PRINT TAB(11) USING UI$;T-TC+1;
1770    PRINT "/";
1780    PRINT USING UI$;DIG;
1790    PRINT TAB(23) USING UD$;DECL;
1800    PRINT TAB(40) USING UD$;PD
1810 NEXT TC
1820 INPUT "Press ENTER to return to menu ",E$
1830 RETURN
2000 'Subroutine for strt line & decl bal table headings  <<<<<<<<<<
2010 PRINT TAB(31) "Book value at"
2020 PRINT "year";TAB(13)"Depreciation";TAB(31)"end of year"
2030 PRINT TAB(6) "Initial value -- $";
2040 PRINT USING UD$;PURCH-SALVAGE
2050 RETURN
```

5, the second is 4, and so forth. Net amount of original price minus salvage is multiplied by the factor each year to calculate dollars of depreciation.

Example 8.15

Depreciation of a $22,000 investment with a six year lifetime and a $4,000 salvage value will show that this method, like the straight-line method, works down to a book value of zero.

The denominator is $1 + 2 + 3 + 4 + 5 + 6$, or 21, and therefore the depreciation factors are 6/21, 5/21, 4/21, and so forth. These factors will multiply $18,000 ($22,000-$4,000) each year. The results are shown in Table 8-3.

COMPUTER PROGRAM FOR DEPRECIATION

The computer program listed in Figure 8-3 prints depreciation schedules as shown in the examples.

Line number 1420 in the declining balance subroutine rounds the factor to four decimal places because most people would use that level of accuracy on their calculators. In the example, 18.8 percent was used, which becomes 0.1818 (a four-decimal-place number) in calculations. If you prefer some other number of decimal places, change the 4 to your number (twice) in line 1420.

Line number 1740 in the sum-of-years subroutine changes the remaining book value to zero when it drops to less than half a cent. This change is just to give the table a neater appearance; otherwise the computer might show a final balance of 1/10,000 of a cent.

Enter the program into your computer and then test it with the examples in this section. The program is structured with a subroutine for each type of depreciation, so you can easily locate any series of lines copied incorrectly.

CHAPTER 9

Techniques of Linear Programming for Analyzing Production Choices

There are many ways to define an optimum arrangement. It might be the arrangement that results in the lowest total cost, lowest labor cost, shortest shipping distance, least overtime, least material, or highest profit. Money does not have to be involved, even indirectly, in the optimization rule; you could seek the arrangement that puts the least burden on a certain ingot furnance or that takes the least traffic near a secured area. If you wanted to test a new form, you might design a system that requires the largest number of entries on the form.

Linear programming allows you to define optimum and to identify the variables you control. Variables you control are the items on which you can make decisions. You might be able to select which production machines will be used, which plants to ship from, the order in which to manufacture subassemblies, or testing standards.

This chapter looks at two types of linear programming. First is the graphical method, in which optimum is shown by lines you draw on a graph. Then the chapter looks at tableau methods, in which you set up an array of numbers and follow a procedure for moving the numbers around the array in certain patterns until they show the optimum arrangement.

GRAPHICAL LINEAR PROGRAMMING

Let us start with a machine that can make either nuts or bolts; it can be switched from one to the other at any time. If it makes only nuts, its capacity for a full day is 1,000 nuts. In one whole day it has the capacity to make 500 bolts. These rates are valid for any part of a day so, for example, if the machine made nuts for 1/10 of a day it would make 100 nuts.

Figure 9–1 shows the production possibilities of this machine for one day. Point a shows that the machine can make 500 bolts if it makes bolts all day, Point c shows that it can also make 800 nuts and 100 bolts in one day. Any combination on the line between points a and b is possible.

It is also possible to produce less than capacity. The machine can make 700

Figure 9–1. Machine can produce any combination in shaded region.

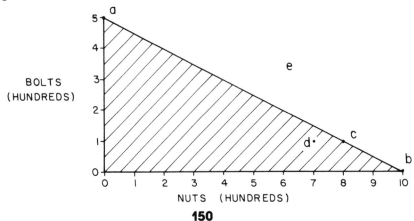

Figure 9–2. Production possibilities for second machine.

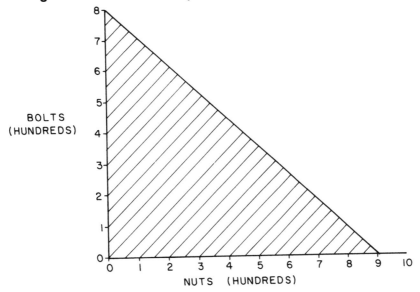

nuts and 100 bolts, point *d*. In fact, the possibilities include all combinations in the shaded region or on the line.

But it is not possible to produce more than the machine's daily capacity. Six hundred nuts and 300 bolts, shown by point *e*, is not a possibility.

Let's look at a process that includes three machines. After the nuts and bolts are made, they go to a plating machine that has a capacity of 900 nuts, 800 bolts, or any combination of those daily rates. Its production possibilities are shown in Figure 9–2.

After plating, the nuts and bolts go to a packaging machine. It can package 1,300 nuts, 400 bolts, or any tradeoff combination in a day.

Figure 9–3 shows all three production possibility lines for this example on one graph. As none of the machines can exceed its capacity, the shaded region of overall possibilities is bounded in all parts by the nearest line. The overall possibilities are those points within or on the boundaries of the polygon whose vertices are *f*, *g*, *h*, *i*, and *j*.

Determining Maximum Profit Possible

Let us now assume that you make a profit of one cent on each nut produced and three cents on each bolt. You now have enough information to determine the maximum profit possible, given the capacity limitations of our machines.

Certainly, profit will not be maximum if you produce some combination that is less than your overall capacity. Therefore, the solution to the maximizing problem will have to be one of the combinations on the bent line *g-h-i-j*, In fact, with very few exceptions, the combination for maximum profit will be one of the vertices *g*, *h*, *i*, or *j*.

Figure 9–3. Production possibilities for three machines are bounded by polygon fghij.

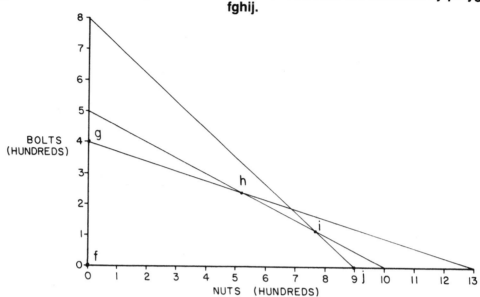

Isoprofit Lines for Indicating Constant Profit

The next step is to draw isoprofit lines—that is, lines of constant profit. For example, you earn \$6 profit by producing 600 nuts. Producing 200 bolts also earns you \$6, as does 300 nuts and 100 bolts. Figure 9–4 shows the \$6 isoprofit line, along with an \$8 isoprofit line.

Notice that the isoprofit lines are parallel. It is a fact that all isoprofit lines for the same profit formula (in this example, one cent for each nut and three

Figure 9–4. Isoprofit lines added to production possibilities.

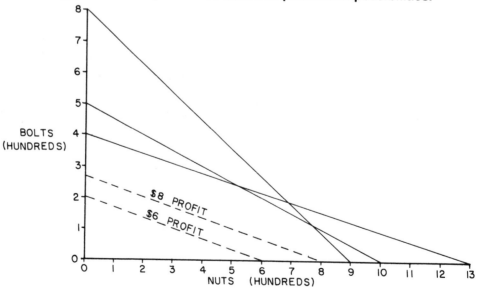

cents for each bolt) are parallel. Therefore, you can continue drawing parallel lines; the line that passes through a vertex farthest from the origin (point *f*) represents maximum profit. The combination of products at that vertex should be produced.

If this graph is drawn very carefully on a large sheet of graph paper with small divisions, a maximum profit of $12.40 will be found at point *h*. The product mix at that point is 520 nuts and 240 bolts. Both the packaging and the initial producing machines will be operating at capacity but there will be some idle time at the plating machine.

The graphical method can include any number of operations. However, the products are limited to two because each additional product would require another dimension to the graph.

This graphical solution does not require that isoprofit lines actually be drawn. As maximum profit will occur at one of the vertices, it is only necessary to calculate the profit at each vertex, including the two on the axes.

Following is a general procedure for graphical linear programming.

1. For each operation, locate the full-capacity production of one product on one axis of the graph, and the full-capacity of the other product on the second axis. Draw a straight line between the two points.

2. Calculate profit (or whatever you are optimizing) at each vertex of the resultant polygon.

3. Select the vertex that results in your definition of optimum.

If the same optimum is found at two vertices, it means that the isoprofit lines are parallel to the line segment that joins those two vertices. Then, every combination along that line segment is equally profitable.

TRANSPORTATION MODEL: USING LINEAR PROGRAMMING TO MINIMIZE SHIPPING COSTS

The simplest of the tableau methods was given this name because its applications often include moving or transporting products. Typically a manager is able to supply each customer from one or more of several plants, and there is a certain shipping cost for each plant-customer combination. Given the capacity of each plant and the needs of each customer, the manager wants to optimize the allocations—that is, arrange shipping for a minimum total cost. The procedure will be explained by way of an example.

Example 9.1

Your firm can ship metallic tubing from four plants, which they abbreviate P1, P2, P3, and P4. Capacities of the plants are: P1, 56 tons; P2, 48 tons; P3, 34 tons; P4, 45 tons. Five customers have placed orders as follows: C1, 33 tons; C2, 24 tons; C3, 30 tons; C4, 46 tons; and C5, 21 tons. You know the shipping costs from each plant to each customer; you now want to arrange the shipments for minimum total shipping costs.

The information is entered on the first tableau, shown in Figure 9–5. An extra customer has been added as a dummy column because customers have ordered only 154 tons of the 183 tons available.

Shipping costs per ton are written in the quarter circles for each combination of plant and customer. They are negative quantities because they are costs and you want to minimize them.

Getting Started: Some Initial Allocations

You have to start with some set of allocations. They can be arbitrary or they can be based on experience and prior arrangements. The closer the initial allocations are to optimum, the fewer steps you will probably need to take.

Let's arbitrarily start in the upper left-hand corner and fill each customer's order from the next plant that has material to ship. For this loading, and throughout the manipulations, it is necessary that the sum in each column be exactly the amount each customer ordered, and the sum in each row be exactly each plant's capacity. The first tableau, with initial loading, is shown in Figure 9–5.

Initial total shipping costs

To follow the changes in shipping costs during manipulations, calculate total shipping costs for this initial set of allocations.

$$S1 = 33(4) + 23(2) + 1(2) + 30(4) + 17(3) + 29(2) + 5(5) + 16(4) + 29(0)$$

$$= \$498$$

Rule for manipulating allocations

Pick an empty box and follow a loop that returns to that box, making right-angle turns at boxes that have assignments. For example, start at the intersection of P3 and C3. (From now on each box will be referred to by the plant and customer that intersect there: P3-C3.) There is a loop consisting of P3-C4, P2-C4, P2-C3, and back to P3-C3. After identifying a loop, determine if moving one unit is beneficial. If it is, move as many as possible.

To determine if a move is beneficial, experiment with the loop while *keeping the sum of each row and each column unchanged.*

Figure 9–5. First tableau of transportation model.

SUPPLIER	CAP'Y	C1	C2	C3	C4	C5	CD
P1	56	-4) 33	-2) 23	-1)	-2)	-3)	-0)
P2	48	-3)	-2) 1	-4) 30	-3) 17	-1)	-0)
P3	34	-1)	-3)	-5)	-2) 29	-5) 5	-0)
P4	45	-3)	-4)	-2)	-2)	-4) 16	-0) 29
		33	24	30	46	21	29

Figure 9–6. Tableau after first set of moves.

SUPPLIER	CAP'Y	C1	C2	C3	C4	C5	CD
		CUSTOMER					
P1	56	-4 32	-2 24	-1	-2	-3	-0
P2	48	-3 1	-2 0	-4 30	-3 17	-1	-0
P3	34	-1	-3	-5	-2 29	-5 5	-0
P4	45	-3	-4	-2	-2	-4 16	-0 29
		33	24	30	46	21	29

Example 9.2

To move in the loop just identified, first move one unit from P2-C3 to P3-C3. That leaves the total in the C3 column unchanged, but decreases the total in the P2 row and increases the total in the P3 row. Therefore, one unit must be moved from P3-C4 to P2-C4 to leave all the sums unchanged.

The effect of that set of moves is a \$1 increase in shipping costs from P2-C3 to P3-C3 and a \$1 increase from P3-C4 to P2-C4. As a move in this loop increases shipping costs, no change should be made in this loop.

Another possibility is the P2-C1, P2-C2, P1-C2, P1-C1 loop. Moving one unit from P2-C2 to P2-C1 increases shipping costs by \$1 and moving from P1-C1 to P1-C2 decreases shipping costs by \$2. There is a net decrease in shipping costs through this loop, so the move should be made, P2-C2 limits the number that can be moved in this loop to one. Figure 9–6 shows the tableau after the first set of moves.

It is informative but not necessary to calculate total shipping costs after each move.

$$S2 = 32(4) + 24(2) + 1(3) + 30(4) + 17(3) + 29(2) + 5(5) + 16(4) + 29(0)$$
$$= \$497$$

Figure 9–7. Tableau after second set of moves.

SUPPLIER	CAP'Y	C1	C2	C3	C4	C5	CD
		CUSTOMER					
P1	56	-4 32	-2 24	-1	-2	-3	-0
P2	48	-3 1	-2 0	-4 30	-3 12	-1 5	-0
P3	34	-1	-3	-5	-2 34	-5 0	-0
P4	45	-3	-4	-2	-2	-4 16	-0 29
		33	24	30	46	21	29

Figure 9–8.   Tableau after third set of moves.

SUPPLIER	CAP'Y	C1	C2	C3	C4	C5	CD
P1	56	-4) 32	-2) 24	-1)	-2)	-3)	-0)
P2	48	-3) 1	-2) 0	-4) 30	-3) 0	-1) 17	-0)
P3	34	-1)	-3)	-5)	-2) 34	-5)	-0)
P4	45	-3)	-4)	-2)	-2) 12	-4) 4	-0) 29
		33	24	30	46	21	29

P2-C5, P2-C4, P3-C4, P3-C5 is another loop to investigate. Moving one unit from P2-C4 to P2-C5 reduces shipping costs by $2. Then, to keep the sums unchanged, one unit must be moved from P3-C5 to P3-C4, resulting in a further decrease of $3. The move should be made, and P3-C5 limits the number to five. Figure 9–7 shows the tableau after the move.

Total shipping costs are now

$$S3 = 32(4) + 24(2) + 1(3) + 30(4) + 12(3) + 5(1) + 34(2) + 16(4) + 29(0)$$

$$= \$472$$

A bigger loop includes P4-C4, P4-C5, P2-C5, and P2-C4. The sums of all rows and columns are left unchanged by moving a unit from P4-C5 to P4-C4 (reducing shipping costs by $2) and from P2-C4 to P2-C5 (reducing shipping costs by another $2). This loop offers a saving, and twelve units can be moved. The result is shown in Figure 9–8.

Total shipping costs after move are

$$S4 = 32(4) + 24(2) + 1(3) + 30(4) + 17(1) + 34(2) + 12(2) + 4(4) + 29(0)$$

$$= \$424$$

The P2-CD, P2-C5, P4-C5, P4-CD loop shows no benefit so it will be left unchanged. A decrease in shipping costs can be made by moving four units through the P4-C3, P4-C5, P2-C5, P2-C3 loop. Then twenty-six units can profitably be moved in the P2-CD, P4-CD, P4-C3, P2-C4 loop.

Figure 9–9 shows the tableau after a few more moves. At this point, shipping costs are down to

$$S5 = 24(2) + 30(1) + 2(0) + 21(1) + 27(0) + 33(1) + 1(2) + 45(2)$$

$$= \$224$$

Using Transportation Model When You Define Optimum as a *Maximum* Instead of a *Minimum*

When the transportation model is used to find an arrangement that gives a *maximum* (we just found a minimum because we were working with costs), the

Figure 9–9. With this arrangement, shipping costs are down to $224.

SUPPLIER	CAP'Y	C1	C2	C3	C4	C5	CD
			CUSTOMER				
P1	56	-4/ 0	-2/ 24	-1/ 30	-2/	-3/	-0/ 2
P2	48	-3/	-2/ 0	-4/ 0	-3/ 0	-1/ 21	-0/ 27
P3	34	-1/ 33	-3/	-5/	-2/ 1	-5/	-0/
P4	45	-3/ 0	-4/	-2/ 0	-2/ 45	-4/ 0	-0/
		33	24	30	46	21	29

numbers in the quarter circles should be positive. Items should then be moved through a loop if the move will cause an *increase*.

THE MODI MODEL: A MORE SYSTEMIZED APPROACH TO EXCHANGING NUMBERS IN THE ARRAY

Model similarities

The MODI model, like the transportation model, is an array of numbers at the intersections of rows and columns. In our examples, the numbers are shipping costs, each row represents a plant, and each column represents a customer. The objective with either model is to arrange the variables (allocate plant output to specific customers in our example) so as to optimize some condition (minimize shipping costs in our example). In both models, a number at an intersection (an occupied box) indicates that a shipment is made from the plant in that row to the customer in that column. No number at an intersection (an empty box) means no shipment for that plant-customer combination.

Model differences

The main difference in the models is that the MODI offers a more systematic way to determine when moving numbers from an occupied box to an empty box provides a better arrangement in terms of your definition of optimum. We will explain the MODI model by working through an example—the same example we used with the transportation model, where the goal is to minimize shipping costs.

Figure 9–10 shows that the tableau has one additional row and column, labeled I (for index). The initial allocations in this example made from the upper left, were arbitrary as with the transportation model. Experience or knowledge of a good arrangement could get you to the optimum arrangement in fewer moves.

Index numbers

Start inserting index numbers by placing a zero in P1's index box. Then, for each box in the P1 row that has an allocation, determine an index number

Figure 9–10. MODI tableau is shown with same initial allocations.

SUPPLIER	CAP'Y	I	C1	C2	C3	C4	C5	CD
			CUSTOMER					
P1	56		-4 / 33	-2 / 23	-1	-2	-3	-0
P2	48		-3	-2 / 1	-4 / 30	-3 / 17	-1	-0
P3	34		-1	-3	-5	-2 / 29	-5 / 5	-0
P4	45		-3	-4	-2	-2	-4 / 16	-0 / 29
			33	24	30	46	21	29

such that the algebraic sum of index numbers equals the shipping cost at their intersection.

For example, there is an allocation at P1-C1, so let's place an index number in the box under C1. It will be a number that when added to the zero in P1's index box, will equal the shipping cost of −4 at P1-C1. That number will be −4, and it is shown in C1's index box in Figure 9–11.

In the P1 row there is also an allocation at P1-C2, so an index number in C2's index box is next. It must be −2 because −2 plus 0 equal the shipping cost (-2) at P1-C2.

There are no more allocations in the P1 row or the C1 column. There is one at P2-C2, so an index number can now be placed in P2's index box. It must be 0 because 0 plus −2 equal the shipping cost at P2-C2 (-2).

Now that P2's index is known, and as there are allocations at P2-C3 and P2-C4, index numbers can be calculated for C3 and C4. Then the allocation at P3-C4 means an index number should be calculated for P3. Figure 9–11 is the complete first tableau.

Determining Where to Move

Check an empty box to see if it is beneficial to move into it. Pick an empty box that can be part of a loop, as with the transportation model. Algebraically add the index numbers of the row and column that intersect at that box. If their sum is less than the shipping cost for that box, it will be beneficial to move units into it.

Example 9.3

P2-C1 is empty and in a loop with P2-C2, P1-C2, and P1-C1. The algebraic sum of the index numbers at P2 (0) and C1 (−4) is −4. As −4 is smaller than −3 (the shipping cost at P2-C1), it is beneficial to move into that box. The largest quantity that can be moved in that loop is one. After the move the body of the tableau will be the same as Figure 9–6.

Figure 9–11. Initial allocations and index numbers.

SUPPLIER	CAP'Y	I	C1	C2	C3	C4	C5	CD
			CUSTOMER					
			-4	-2	-4	-3	-6	-2
P1	56	0	-4/ 33	-2/ 23	-1/	-2/	-3/	-0/
P2	48	0	-3/	-2/ 1	-4/ 30	-3/ 17	-1/	-0/
P3	34	1	-1/	-3/	-5/	-2/ 29	-5/ 5	-0/
P4	45	2	-3/	-4/	-2/	-2/	-4/ 16	-0/ 29
			33	24	30	46	21	29

What to Do After Moving Units Through a Loop

Making a move changes one occupied box to empty and one empty box to occupied. Therefore it is necessary to recalculate the index numbers, always starting with zero at P1. Then check empty boxes again and move units if indicated by the algebraic sum of index numbers. Repeat the entire procedure until arriving at a tableau that cannot be improved. For this example, it should have the same allocations as Figure 9–9.

How to Use the MODI Model to Your Advantage

Having to calculate index numbers repeatedly discourages some people from using the MODI model. However, those who have used it usually consider there is a net saving in time because the MODI makes it clearer when a further improvement is possible. Many transportation models have been stopped short of optimum because the analyst didn't see another move that could have been made.

One way to make the MODI easier to work is to draw one tableau in ink. Fill in everything except allocations and index numbers in ink. Write allocations in pencil and then calculate index numbers and write them in pencil. To move quantities in a loop, erase them and make the changes in pencil. Then erase the index numbers and write the new ones in pencil.

Another method is to do the ink work in the preceding paragraph and then make reproductions on the office copier. Use a new copy after moving through a loop. The advantage to this method is that you have a copy of each tableau.

Using MODI Model When Your Definition of Optimum *Maximizes* Instead of *Minimizes* Some Condition

The quantity being optimized can be a desirable item, such as profit, instead of a cost. Then the numbers in the quarter circles should be positive, and a *maximum* is the objective. Units should be moved into an empty box when the algebraic sum of index numbers is *larger* than the number in the quarter circle.

CHAPTER 10

Queuing Shortcuts: Practical Applications for Speeding up People, Machines, and Materials

Employees are not productive when they are waiting in lines. Machines and other equipment are not productive when they are waiting to be serviced. As both types of waiting are costly, there has been a considerable amount of study on "*queuing theory*."

Queue, which refers to forming or waiting in a line, is one word that did not become a regular part of American English until recently. It was always *waiting line theory* in the United States, but now we more often hear *queuing theory*.

From formulas that predict the length of lines at toll booths and carnival rides, management science picked up the theory to predict lines at copy machines, tool counters, and company cafeterias.

But the queue does not have to consist of people and cars physically getting in a line; it can also be a list. With that thought, the field is open to consider that machines that break down are in line waiting for service. Incoming telephone calls put on hold "until the next available operator," assembly line rejects backed up at a repair facility, and cars arriving at a toll booth can be examined through the same formulas. This chapter explores queuing theory's applications to production management.

Also included in this chapter is a special section on simulation, to show you how to simulate a period of time and observe the line formation. Manual and computer simulations are given.

Symbols used

As the same symbols are used throughout this chapter, they are all explained here instead of repeating them by each formula.

c = the number of service channels

f = utility factor; ratio of arrival rate to service rate—must be less than one

n = number of units being considered

n_s = service rate—average number of units served per unit of time

= reciprocal of time to serve one unit

P_n = probability of n units being in the system—includes the unit(s) being served as well as those in the queue

q = average number of units in the queue

q_t = average number of units in the system—counts those being served as well as those in the queue

r_a = average number of arrivals per unit of time

s = number of units in the source, when finite source is assumed

t_q = average time a unit spends in the queue

t_t = average time a unit spends in the system—counts time in the queue as well as time being served

QUEUING THEORY VOCABULARY

To make it easier to transfer the theory from one application to another, this chapter uses the universal vocabulary of queuing theory. Thus, *arrival* refers to anything getting on the waiting line or list. Incoming telephone calls, workers going to the nurse's, orders placed in the copy facility's in-basket, and production machine breakdowns are all arrivals. Other universally recognized words and phrases are used in this chapter. For example, the *service facility* is that for which the arrivals are queuing — it may be answering the telephone, caring for injured workers, using the copying machine, or repairing a production machine. Sometimes a service facility comprises more than one *service channel*, as when there are two nurses and the next one to become free takes the next person in the queue. Formulas in this chapter refer to either *single channel* service or *multiple channel* service.

Figure 10-1. Natural distribution of arrival rates.

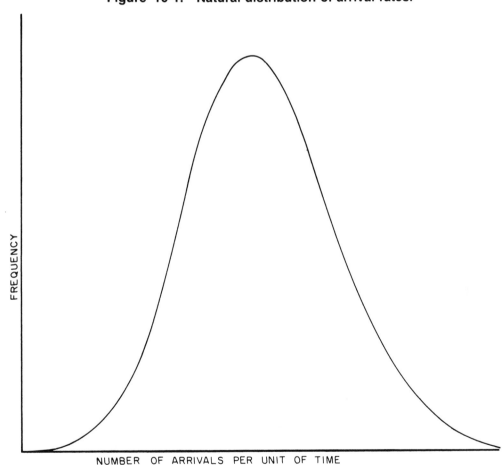

TIMING OF ARRIVALS

The basis for queuing theory is that, although the timing between individual arrivals is random, their averages are predictable. Unless there is an overriding influence (such as a traffic light that bunches traffic before the toll booth, or an electric surge that burns out lightbulbs), arrivals can be expected to fit the distribution shown in Figure 10-1. For example, Company A kept records of the number of lightbulbs that burned out in the production area each day. Table 10-1 shows the tally for 129 days, with the first ten days on the first line, the next ten days on the second line, and so forth.

Table 10-2 summarizes the data in Table 10-1. It shows, for example, that on three days of the 129 recorded there were no lightbulb failures (first line of Table 10-2), on two days there were two failures each (second line), and so forth.

Figure 10-2 is a graph of the same data. Its general shape is similar to Figure 10-1, although you can note some differences (besides the fact that Figure 10-1 is rotated ninety degrees). If you enlarged the sample from 129 to several thousand, it would more closely resemble the theoretical distribution.

Another example: Company B uses a dumbwaiter to move parts between floors of the production department. An analyst timed how many groups of parts arrived at the dumbwaiter every twelve minutes, and the result is shown in Table 10-3. The table's first row shows the first ten recordings of twelve-minute intervals, the second row shows records eleven through twenty, and so on, for a total of fifty-two intervals.

Again, the results are summarized into a frequency table, Table 10-4, and graphed in Figure 10-3. The first line of Table 10-3 shows that there was one twelve-minute interval in which no groups of parts arrived; the fourth line shows that there were twelve intervals in which three groups of parts arrived.

The graphs are not exact copies of Figure 10-1, but they are similar. In general, the larger the number of arrivals you record, the more you can expect the graph to look like the theoretical distribution shown in Figure 10-1.

Table 10-1.   Daily lightbulb failures for 129 days.

6	10	11	13	13	9	5	5	10	7
18	7	9	12	3	3	8	4	0	11
4	5	22	19	9	9	7	3	5	10
7	5	4	0	8	6	4	6	14	8
22	2	14	12	9	14	13	3	8	4
19	14	12	11	12	13	2	16	7	3
6	7	4	15	6	11	9	3	14	7
11	16	11	0	19	6	12	4	11	7
12	8	4	5	8	7	12	14	15	8
10	6	11	6	5	12	11	9	15	3
12	11	13	10	10	5	7	4	4	14
15	16	5	19	10	16	9	6	18	4
17	6	3	13	5	10	9	15	6	

Table 10-2. Summary of the data in Table 10-1.

Item number	Failures per day	Frequency
1	0	3
2	2	2
3	3	8
4	4	11
5	5	10
6	6	11
7	7	10
8	8	7
9	9	9
10	10	8
11	11	10
12	12	9
13	13	6
14	14	7
15	15	5
16	16	4
17	17	1
18	18	2
19	19	4
20	22	2

Here is one more example of arrivals, because the concept of arrivals per unit of time is important for understanding queuing theory. Table 10-5 shows machine breakdowns per day for fifty-six days. The frequency summary in Table 10-6 shows how often each number of breakdowns occurred, and Figure 10-4 is a graph of the frequency summary.

Figure 10-2. Bar graph of data in Table 10-2.

Each asterisk represents .5

```
 0 to   1   ******
 1 to   3   ***************
 3 to   5   *********************************************
 5 to   7   *********************************************
 7 to   9   *********************************
 9 to  11   ***********************************
11 to  13   ******************************
13 to  15   ************************
15 to  17   **********
17 to  19   ************
19 to  21
21 to  23   ****
```

Table 10-3. Parts arriving at dumbwaiter in twelve minutes.

9	6	3	7	4	3	5	5	2	4
5	3	8	5	3	4	3	4	1	4
4	4	2	4	8	3	0	3	5	4
6	2	3	4	7	2	8	1	5	2
2	4	3	3	7	5	3	5	7	8
6	3								

SERVICE TIMES

Negative exponential service time

Formulas in this chapter are grouped according to service time distributions. A considerable amount of empirical data has shown researchers that the length of time required in nearly all service facilities is described by an equation with a negative exponential term. For any given mean value of time, this type of equation shows a higher probability for smaller values, and the probability of larger values drops off rapidly.

Constant service time

Another distribution used, and one that is easier to picture, is constant service time, where the same amount of time is spent on every unit that comes through the queue. For example, suppose several branches of a conveyor system converge at a point that reads a sixteen-bar code on every container the system carries. As the computer takes the same length of time to read the code on each container, constant service time applies.

Poisson service time

The third distribution considered in this chapter is called a Poisson distribution. It is characterized by low probability for low and high lengths of time, with the highest probability near the mean value.

Table 10-4. Summary of the data in Table 10-3.

Item number	Number of groups	Frequency
1	0	1
2	1	2
3	2	6
4	3	12
5	4	11
6	5	8
7	6	3
8	7	4
9	8	4
10	9	1

Figure 10-3.   Bar graph of data in Table 10-4.
Each asterisk represents .5

```
             0    **
    0 to     1    ****
    1 to     2    ************
    2 to     3    *************************
    3 to     4    ***********************
    4 to     5    ******************
    5 to     6    ******
    6 to     7    ********
    7 to     8    ********
    8 to     9    **
```

Selecting the service time

Unless you are certain that one of the last two distributions applies, use the formulas that are based on negative exponential distribution. Actually, this selection is not a critical one, because it changes only the absolute numbers in your answers (number waiting, length of wait). Investigators are often more interested in *changes* in these numbers as they experiment with different values of the variables.

Due to space limitations, this book does not present every combination of arrival and service time distributions. Neither does it include every formula for the situations that are included. The most likely situations have been selected and the most meaningful formulas have been made available for each situation to present, explain, and demonstrate.

CONDITIONS (ASSUMED AND STATED) FOR SOURCE

Formulas in the following pages are grouped according to arrival rate and service time distributions. One other condition is the size of the source of arrivals. Sources are described here as either finite or infinite.

Finite versus infinite sources

A finite source is treated differently because the percentage changes when each unit leaves. For example, if there are five units in the source, and one of

Table 10-5.   Machine breakdowns per day.

3	7	7	0	0	6	6	7	9	3
4	9	6	10	3	5	3	5	6	5
4	5	8	2	7	5	3	4	4	1
1	3	10	5	5	8	3	7	2	4
3	3	2	5	5	4	7	8	3	11
5	2	7	4	5	4				

Table 10-6. Summary of the data in Table 10-5.

Item number	Number of breakdowns	Frequency
1	0	2
2	1	2
3	2	4
4	3	10
5	4	8
6	5	11
7	6	4
8	7	7
9	8	3
10	9	2
11	10	2
12	11	1

them leaves, twenty percent of them have left. When the next one leaves, twenty-five percent of them will be leaving. One unit leaving from a source of 100 represents 1.00 percent; it is almost the same when the next one leaves—1.01 percent. Therefore, applying infinity formulas to a finite source will affect the results only slightly unless the number remaining is very small. The results of using finite source formulas tend to approach the results of assuming infinite source as the number in the source increases. To calculate general working figures, many investigators always use the formulas for infinite source. They're easier, and the results are useful for planning.

The size of the source has one other effect. If arrivals come from a source of five, then the size of the queue is limited to four (one in service plus four on line), even if every one from the source needs service at the same time. When you assume an infinite source, you also assume no limit to the size of the queue.

Figure 10-4. Bar graph of data in Table 10-6.

Each asterisk represent .5

```
 0    ****
 1    ****
 2    ********
 3    ******************
 4    ****************
 5    ********************
 6    ********
 7    **************
 8    ******
 9    ****
10    ****
11    **
```

One condition can arise from using formulas for an infinite source, when the actual source is small. The formulas could tell us that there are more units in the queue than were available to arrive for service. This is usually not a problem because the investigator can simply disregard any impossible answers. Even more important, the investigator usually stops when he or she sees that the system results in a growing queue.

Variables to Consider in a Queuing Study

Purpose of investigation

The last sentence in the previous paragraph brings up an important point. The main variable in a queuing study is usually service time. You generally can't tell the machines to break down less often, although the purpose of a study might be to find an optimum maintenance program. Then the rate of breakdown, used as arrival rate, would be a controllable variable. Typically an investigator assumes certain arrival and service rates, makes the calculations, and finds that the queue would grow during a shift. Then he or she experiments with the service rate until he or she finds an acceptable set of conditions.

One rule that always applies—average service rate must be faster than average arrival rate. Otherwise, even if the rates are the same, the queue will grow without bound.

Other assumptions

To avoid undue complications, assume no priorities (no cuts allowed). In other words, units are taken for service in the order in which they arrive. When working with multiple service facilities, assume that there is a single queue and the next available service facility takes the first unit in the queue, similar to the system banks now use. You do not have a queue forming at each service facility.

Finally, assume that every unit arriving either goes directly into service or gets on the queue. There is no third choice, and no unit turns aways because the queue is too long.

INFINITE SOURCE

All the formulas that are usually considered appropriate for hand calculation assume an infinite source. Formulas and examples for finite source are given after this section. However, most of the practical examples using finite source will necessarily be covered after computer solutions are discussed.

Negative Exponential Service Time

This is the most common situation. It is also often used as the default situation when no specific service time distribution is identified.

Single channel

The phrase "single channel" describes the situation in which there is just one service facility to handle all units. The single service facility may be one mechanic, one tool counter, or one copying machine. It could also be a team of mechanics. The important point is that each arrival goes to the same service facility, and only one can be served at a time.

Because these formulas are simpler than those for multiple channels, their use is preferable when there is a choice. For example, if the reproduction facility has many copying machines, you might think of the entire facility as a single channel if you can apply a single average service rate to it. Likewise, if repairs are handled by a group of mechanics working together, the group is a single service facility or channel.

$$q = \frac{r_a^2}{n_s(n_s - r_a)}$$

$$q_t = \frac{r_a}{n_s - r_a}$$

$$t_q = \frac{r_a}{n_s(n_s - r_a)}$$

$$t_t = \frac{1}{n_s - r_a}$$

$$f = \frac{r_a}{n_s}$$

$$P_o = 1 - f$$

$$P_n = f^n P_o$$

Example 10.1

Assembly line defects average six per hour. Defective units are repaired by a technician who averages eight repairs per hour. As one step in examining the procedure's efficiency, you want to study all the numbers relating to queuing.

Defects are arrivals, so r_a is six per hour. Service rate, n_s, is eight per hour. As the service rate is larger than the arrival rate, you can proceed to use the formulas. Continuing operation of the assembly line provides an infinite source.

$$q = \frac{6^2}{8(8 - 6)}$$

$$= 2.25$$

On the average, there are 2.25 completed units from the assembly line that cannot be shipped until the technician works on them.

$$q_t = \frac{6}{8 - 6}$$

$$= 3$$

On the average, three completed units are held up, including the one the technician is working on.

$$t_q = \frac{6}{8(8-6)}$$

$$= 0.375$$

On the average, once a unit is removed from the assembly line, it is held up 0.375 hour ($22\frac{1}{2}$ minutes) until the technician can work on it.

$$t_t = \frac{1}{8-6}$$

$$= 0.5$$

On the average, a defective unit is delayed 0.5 hour before its problem is corrected and it can be shipped.

$$P_o = 1 - \frac{6}{8}$$

$$= 0.25$$

There are no units in the system (the repair system, off the assembly line), and therefore the technician is idle twenty-five percent of the time. Probabilities of other quantities being in the system are calculated in the following table.

Number of units	Probability of that number of units being in the system
1	$(6/8)^1(0.25) = 0.19$
2	$(6/8)^2(0.25) = 0.14$
3	$(6/8)^3(0.25) = 0.11$
4	$(6/8)^4(0.25) = 0.08$
5	$(6/8)^5(0.25) = 0.06$

With these numbers the production executive can objectively consider various actions. For example, the technician might be given work to do on a fill-in basis, or rejects might be allowed to accumulate to a certain number and then the technician can work through them. That part of the investigation depends on the cost of holding a completed unit.

Another possibility, just the opposite of allowing rejects to accumulate, would be having a substitute technician trained for times when the regular technician is absent.

Multiple Channels

Now let's work with infinite source, negative exponential service time, Poisson arrival rates, and multiple channels. The last condition means that more than one service facility is fed from a single queue. When a unit in the queue reaches the front of the line, it is taken by the next available service facility.

The probability of all the service facilities being idle (probability of zero units in the system) is given first because it is used in the other equations and therefore must be calculated first.

$$P_o = \cfrac{1}{\left[\displaystyle\sum_{n=0}^{c-1} \frac{1}{n!}\left(\frac{r_a}{n_s}\right)^n\right] + \dfrac{1}{c!}\left(\dfrac{r_a}{n_s}\right)^c \dfrac{cn_s}{cn_s + r_a}}$$

$$q = \cfrac{r_a\, n_s \left(\dfrac{r_a}{n_s}\right)^c}{(c-1)!\,(cn_s - r_a)^2}\, P_o$$

$$q_t = q + \frac{r_a\, c}{n_s}$$

$$t_q = \cfrac{n_s \left(\dfrac{r_a}{n_s}\right)^c}{(c-1)!\,(cn_s - r_a)^2}\, P_o$$

$$t_t = t_q + \frac{c}{n_s}$$

$$= \frac{q_t}{r_a}$$

$$f = \frac{r_a}{cn_s}$$

Calculating P_O must be done very carefully because it involves summation, factorial, and exponents. In addition, it is a critical calculation because an error there will affect all the other values. The computer program given later in this chapter should be welcomed by anyone who has worked these formulas manually, even with a calculator.

Example 10.2

A production area has a large number of coil-winding machines that break down on an average of five per day. Three mechanics work separately (each is a service channel) and each averages half a day to place a machine back in operation. Solve all the formulas.

Notice that you first have to calculate the service rate; you are given its reciprocal, service time.

$$\text{rate} = \frac{1}{\frac{1}{2} \text{ per day}}$$

$$= 2 \text{ per day}$$

Each mechanic averages two machines returned to operation daily, so the system service rate, n_s, is six per day.

$$P_o = \frac{1}{\left[\displaystyle\sum_{n=0}^{3-1} \frac{1}{n!}\left(\frac{5}{6}\right)^n\right] + \frac{1}{3!}\left(\frac{5}{6}\right)^3 \frac{3(6)}{3(6)-5}}$$

$$= 0.4321$$

There is a forty-three-percent probability of every machine being operational and all the mechanics being idle.

$$q = \frac{5(6)\left(\dfrac{5}{6}\right)^3 (0.4321)}{2![3(6)-5]^2}$$

$$= 0.0222$$

On the average, very few machines are broken down and waiting for a mechanic. This small number is reasonable because five break down on a typical day, and the mechanics normally repair six.

$$q_t = 0.0222 + \frac{2(5)}{6}$$

$$= 1.6889$$

This production area averages more than one machine out of operation, either being repaired or waiting for a mechanic.

$$t_q = \frac{6\left(\dfrac{5}{6}\right)^3 0.4321}{2![3(6)-5]^2}$$

$$= 0.0044 \text{ day}$$

$$= 0.0355 \text{ hour}$$

After breaking down, an average machine waits a very short time before a mechanic begins working on it. This is consistent with the previous finding that there is seldom a waiting line. It is assumed here that a day means eight hours.

$$t_t = 0.0044 + \tfrac{3}{6}$$

$$= 0.5044 \text{ day}$$

Because there is seldom a waiting line, a disabled machine can be expected back in operation in just about the time it takes a mechanic to repair it.

Constant Service Time

You are still looking at infinite source and Poisson arrival distribution. Now consider that the same service is to be performed on every unit when it reaches the service facility. This means that n_s is a constant. For example, to speed workers through the company cafeteria, a booth sells $5.00 coupon books to those who do not need change.

Single Channel With Constant Service Time

Only the single service facility arrangement will be examined with constant service time. If there should be multiple service channels, it might be possible to treat them as a single channel. Other possibilities include using negative exponential service time as an approximation, or using simulation as described later in this chapter.

$$q = \frac{r_a{}^2}{2n_s(n_s - r_a)}$$

$$q_t = q + \frac{r_a}{n_s}$$

$$t_q = \frac{r_a}{2n_s(n_s - r_a)}$$

$$t_t = t_q + \frac{1}{r_a}$$

Example 10.3

A company makes so many single copies of $8\frac{1}{2} \times 11$ inch loose sheets, printed on one side, that it sets up a separate copying machine for them. It operates at a constant 400 copies per hour. Sheets to be copied arrive according to a Poisson distribution with a mean value of 398 per hour. Solve all the formulas.

$$q = \frac{398^2}{2(400)(400 - 398)}$$

$$= 99$$

Because the arrival rate is only slightly less than the service rate, an arrival will usually find others waiting.

$$q_t = 99 + \frac{398}{400}$$

$$= 100$$

Adding the number being served to those waiting to be served increases slightly the number in the system.

$$t_q = \frac{398}{2(400)^2}$$

$$= 0.2488 \text{ hour}$$

$$= 14.93 \text{ minutes}$$

Each of the ninety-nine units in a typical queue waits an average of just under fifteen minutes to get to the copying machine.

$$t_t = 0.2488 + \frac{1}{398}$$

$$= 0.2513 \text{ hour}$$

$$= 15.08 \text{ minutes}$$

On the average a copied sheet is ready just over fifteen minutes after being put in.

Poisson Service Time

A fairly common situation finds arrival rates and service rates having the same distribution patterns. It still holds that the queue will grow without bound unless the mean arrival rate is less than the mean service rate (f must be less than one).

Single Channel

The arrangment seen most often, and therefore the one covered here, has a single service facility serving all arrivals.

$$q = \frac{r_a^2}{n_s(n_s - r_a)}$$

$$q_t = \frac{r_a}{n_s - r_a}$$

$$t_q = \frac{r_a}{n_s(n_s - r_a)}$$

$$t_t = \frac{1}{n_s - r_a}$$

$$P_n = \left(1 - \frac{r_a}{n_s}\right)\left(\frac{r_a}{n_s}\right)^n$$

Example 10.4

Throughout the shift, production workers go to the stock counter at an average rate of fifteen per hour. There the clerk dispenses expendable equipment and small hardware items, at an average rate of one every $3\frac{1}{2}$ minutes. Observations show that the dispensing time distribution has a Poisson shape. Solve all the formulas to get numbers you will use in a report to top management.

As n_s used in the formulas is service *rate*, and you were given its reciprocal, service *time*, you must first convert to rate. At the same time you will make the units consistent—multiply by 60 to give a per-hour rate.

$$n_s = \frac{1}{3\frac{1}{2}}(60)$$

$$= 17.14 \text{ per hour}$$

This is the mean service rate to be used in the following calculations.

$$q = \frac{15^2}{17.14\,(17.14 - 15)}$$

$$= 6.13$$

On the average, there are always more than six production workers waiting on line at the stock counter.

$$q_t = \frac{15}{17.14 - 15}$$

$$= 7.01$$

About seven workers are usually away from productive work, either on line or being served.

$$t_q = \frac{6.13}{15}$$

$$= 0.41 \text{ hour}$$

$$= 24.52 \text{ minutes}$$

Typically an individual waits nearly twenty-five minutes to get to the counter.

$$t_t = \frac{7.01}{15}$$

$$= 0.47 \text{ hour}$$

$$= 28.04 \text{ minutes}$$

Including time at the counter, each employee who needs its services can expect to spend twenty-eight minutes away from regular work.

The following table calculates the probability of different numbers of workers being in the system (being on line or being served).

Number of units	Probability of that number of units being in the system.
0	$(1-15/17.14)(15/17.14)^0 = 0.1249$
1	$(1-15/17.14)(15/17.14)^1 = 0.1093$
2	$(1-15/17.14)(15/17.14)^2 = 0.0956$
3	$(1-15/17.14)(15/17.14)^3 = 0.0837$
4	$(1-15/17.14)(15/17.14)^4 = 0.0732$
5	$(1-15/17.14)(15/17.14)^5 = 0.0641$

FINITE SOURCE

When the source of arrivals consists of a fixed quantity, the formulas quickly become difficult to manage. Let's take just a short look at them; later, the computer program will give complete solutions.

Single Channel

Although the following formulas apply specifically to Poisson arrivals and negative exponential service times, they are a good approximation for other situations.

$$P_n = \frac{s!}{(s-n)!} \left(\frac{r_a}{n_s}\right)^n P_0$$

$$q = \sum_{n=1}^{s} (n-1) P_n$$

$$t_q = \frac{q}{r_a}$$

These formulas are much less friendly than they appear at first. They involve factorials, exponents, and summations. Worse than that, P_0 is not given explicitly; it must be calculated as follows.

Calculating P_0

As the arrivals come from a finite source of s, we know that the sum of probabilities of each number must add to one. In other words, if there are three units in the source, then the only possibilities are that there are zero, one, two, or three units in the system. (In the system means on the waiting line or receiving service.) Therefore the probability of zero units in the system, plus the probability of one, plus the probability of two, plus the probability of three must add up to one.

Using that fact, calculate P_0 for a factory in which r_a is 0.8 and n_s is 1.0 (r_a/n_s is 0.8) and there are three units in the source. Such a small source is used to keep the example simple. From the work involved, you will see why analysts turn to computer solutions. Practical situations usually have twenty, fifty, or more units in the source, and manual calculation is unthinkable for most. Adding to the problem is the fact that any error in calculating P_0 will make all the other calculations wrong.

$$P_0 = 1.0\, P_0$$

$$P_1 = \frac{3!}{(3-1)!}\,(0.8)^1 P_0 = 2.4\, P_0$$

$$P_2 = \frac{3!}{(3-2)!}\,(0.8)^2 P_0 = 3.84\, P_0$$

$$P_3 = \frac{3!}{(3-3)!}\,(0.8)^3 P_0 = 3.07\, P_0$$

Adding the last numbers in each row gives $10.31\, P_0$, which must equal one because it accounts for every possibility.

$$10.31\, P_0 = 1$$

Therefore

$$P_0 = .097$$

Now you can calculate a probability table.

Number of units	Probability of that number of units being in the system
0	$= 0.097$
1	$\dfrac{3!}{(3-1)!}\,(0.8)^1\,(0.097) = 0.233$
2	$\dfrac{3!}{(3-2)!}\,(0.8)^2\,(0.097) = 0.373$
3	$\dfrac{3!}{(3-3)!}\,(0.8)^3\,(0.097) = 0.298$

Use the summation formula to find average number in the queue.

$$q = (1-1)(0.233) + (2-1)(0.373) + (3-1)(0.298)$$

$$= 0.9690$$

There will usually be about one unit waiting.

$$t_q = \frac{0.9690}{0.8}$$

$$= 1.2113$$

On the average, a unit going for service will spend 1.21 time units waiting.

Again, these formulas will be included in the computer program. Do not be concerned about working with them unless you enjoy this kind of mathematics as recreation.

Multiple Channels

You are still working with a finite source, Poisson arrival rate, and negative exponential service time. Now there are several service facilities; as each becomes available it will take the next unit that is waiting in the queue. First you define a provisional P_n, which will become part of the expression for P_n.

$$PP_n = \frac{s!}{(s-n)!n!} \left(\frac{r_a}{n_s}\right)^n P_0$$

Then the formulas are

$$P_n = PP_n \qquad \text{when } o \leq n \leq c$$

$$P_n = \left(\frac{n!}{c!c^{n-c}}\right) PP_n \qquad \text{when } c \leq n \leq s$$

$$q = \sum_{n=c}^{s} (n-c) P_n$$

$$t_q = \frac{q}{r_a}$$

Calculating P_0 without a computer becomes even more involved for this multiple channel situation because of switching formulas along the way. Notice that one probability formula applies when there is a service channel for each arrival. Then switch to the other formula to determine probabilities of more units than there are service channels. Let's leave the demonstration of this one for the computer program.

COMPUTER PROGRAM FOR QUEUING CALCULATIONS

A computer program that will solve these queuing formulas is included for two reasons. First, some of the formulas become quite involved, even when you use a calculator. Second, with the computer doing all the work, you are encouraged to experiment. There is no limit to the changes you can try just to see the effect.

The program listed in Figure 10-5 has been thoroughly tested on an IBM PC. Few, if any, changes will be required to run the program on other computers, because only non-specialized BASIC commands are used. The program explanation on the following pages includes details to allow you to decide whether you can use the program directly, or whether some adaptations will be needed. There is also liberal use of remarks in the program lines (beginning with the apostrophes) to help you identify what each part of the program does. If you are

Figure 10-5. Computer program for queuing calculations.

```
10 '***********************************************************************
20 '** QUEUE.BAS    Queuing solutions for Handbook of Manufacturing  **
30 '** and Production Management Formulas, Charts, and Tables         **
40 '***********************************************************************
50 CLEAR,,,32768!
60 COLOR 3,4
70 DIM P(100),T(100)
80 CLS
90 PRINT
100 PRINT TAB(25) "INFINITE SOURCE"
110 PRINT TAB(10) "Negative exponential service time"
120 PRINT " 1  Single channel"
130 PRINT " 2  Multiple channels"
140 PRINT: PRINT TAB(10) "Constant service time"
150 PRINT " 3  Single channel"
160 PRINT: PRINT TAB(10) "Poisson service time"
170 PRINT " 4  Single channel"
180 PRINT: PRINT TAB(26) "FINITE SOURCE"
190 PRINT TAB(10) "Negative exponential service time"
200 PRINT " 5  Single channel"
210 PRINT " 6  Multiple channels"
220 PRINT :INPUT "Select by number ",S
230 IF S>0 AND S<7 THEN 260
240 PRINT "Select between 1 and 6 please!"
250   GOTO 90
260 CLS
270 PRINT "What is the basic unit of"
280 INPUT "time (hour, day, minute, etc.) ";U$
290 PRINT: INPUT "Mean arrival rate ";AR
300 PRINT: PRINT "Mean number served per ";U$;
310 INPUT " ",NS
320 ON S GOTO 340,470,690,780,900,960
330 '(1) Infinite source, neg exp serv time, single chan  <<<<<<<<<<<<
340 GOSUB 4010   ''''''''''COMMON CALCULATIONS
350 GOSUB 4110   ''''''''''DISPLAY CALCULATED VALUES
360 PT$="probability table"
370 GOSUB 4310   ''''''''''ASK ABOUT TABLE AND RANGE
380 IF T$="R" THEN 80
390 PO=1-AR/NS
400 FOR N=SU TO LU
410    PQ=((AR/NS)^N)*PO
420    PRINT TAB(6)N; TAB(27)PQ
430 NEXT N
440 PT$="more probabilities"
450 GOTO 370
460 '(2) Infinite source, neg exp serv time, multiple chan  <<<<<<<<<<<<
470 PRINT: INPUT "How many channels ";CH
480 PRINT
490 T1=0
500 FOR CC=0 TO CH-1
510    F=CC
520    GOSUB 4610   ''''''''''FACTORIAL
530    T1=T1+(1/FF#)*(AR/NS)^CC
540 NEXT CC
550 F=CH
560 GOSUB 4610   ''''''''''FACTORIAL
570 T2=(1/FF#)*(AR/NS)^CH*CH*NS/(CH*NS-AR)
580 PO=1/(T1+T2)
590 QN=NS*PO*(AR/NS)^CH
600 QD=(FF#/(C-1))*(CH*NS-AR)*(CH*NS-AR)
```

Figure 10-5. (continued)

```
610 Q=QN*AR/QD
620 QT=Q+CH*AR/NS
630 TQ=QN/QD
640 TT=TQ+CH/NS
650 GOSUB 4110  '''''''''''''DISPLAY RESULTS
660 INPUT "Press <Enter> for main selection list ",E$
670 GOTO 80
680 '(3) Inf source, const serv time, single chan  <<<<<<<<<<<<
690 GOSUB 4010  '''''''''''COMMON CALCULTIONS
700 Q=Q/2
710 TQ=TQ/2
720 QT=Q+AR/NS
730 TT=TQ+1/AR
740 GOSUB 4110  '''''''''''''''''''''DISPLAY RESULTS
750 INPUT "Press <Enter> for main selection list ",E$
760 GOTO 80
770 '(4) Inf source, Poisson serv time, single chan  <<<<<<<<<<<<<<
780 GOSUB 4010  '''''''''''''COMMON CALCULATIONS
790 GOSUB 4110  '''''''''''DISPLAY RESULTS
800 PT$="probability table"
810 GOSUB 4310  '''''''''''ASK ABOUT PROB TABLE
820 IF T$="R" THEN 80
830 FOR N=SU TO LU
840    PQ=(1-AR/NS)*(AR/NS)^N
850    PRINT TAB(6)N; TAB(27)PQ
860 NEXT N
870 PT$="more probabilities"
880 GOTO 810
890 '(5) Finite source, neg exp serv time, single chan  <<<<<<<<<<<<
900 GOSUB 4710  ''''''''''FINITE CALCULATIONS
910 GOSUB 4440  '''''''''TABLE HEADINGS
920 GOSUB 5010  ''''''''''FINITE TABLE
930 INPUT "Press <Enter> to return to main selection list ",E$
940 GOTO 80
950 '(6) Finite source, neg exp serv time, multiple chan  <<<<<<<<<<<<
960 PRINT: INPUT"How many channels ";CH
970 GOSUB 4710  ''''''''''FINITE CALCULATIONS
980 GOSUB 4440  '''''''''TABLE HEADINGS
990 GOSUB 5010  ''''''''''FINITE TABLE
1000 INPUT "Press <Enter> for main selection list ",E$
1010 GOTO 80
1020 END
4000 'Subroutine COMCALC makes common calculations  $$$$$$$$$$$$
4010 NR=NS-AR
4020 Q=AR*AR/(NS*NR)
4030 QT=AR/NR
4040 TQ=AR/(NS*NR)
4050 TT=1/NR
4060 RETURN
4100 'Subroutine DISPCALC displays calculated values  $$$$$$$$$$$$
4110 CLS
4120 PRINT"     On the average, a unit will spend";TQ
4130 PRINT U$+"S waiting to be served.  It will spend"
4140 PRINT TT;U$+"S in the system, including the time it"
4150 PRINT"is being served."
4160 PRINT
4170 PRINT"     The average number of units waiting to"
4180 PRINT"be served will be";Q;"and the average number of"
4190 PRINT"units in the system, including those being served,"
4200 PRINT"will be";QT;"."
```

Figure 10-5. (continued)

```
4210 PRINT
4220 RETURN
4300 'Subroutine ASKPROB asks if probability table is wanted   $$$$$$$$$$$
4310 PRINT "Type P to see ";PT$
4320 INPUT "  or R to return to selection list ",T$
4330 IF T$="P" THEN 4370
4332 IF T$="R" THEN 4470
4340 PRINT "Type P or R please!"
4350 PRINT: GOTO 4310
4360 'ask range for probabilities  <<<<<<<<<<<<<
4370 CLS
4380 PRINT"What is the smallest number of units whose"
4390 INPUT"probability is to be shown ";SU
4400 PRINT: INPUT"... and the largest ";LU
4410 IF LU>SU THEN 4440
4420 PRINT"Largest should be larger than smallest"
4430 PRINT :GOTO 4380
4440 PRINT TAB(2)"Number";TAB(14)"Probability of that number being"
4450 PRINT"of units";TAB(20)"in the system"
4460 PRINT STRING$(45,".")
4470 RETURN
4600 'Subroutine FACT! calculates factorial   $$$$$$$$$$$$$$
4610 FF#=1
4620 IF F<2 THEN 4660
4630 FOR C=2 TO F
4640    FF#=FF#*C
4650 NEXT C
4660 RETURN
4700 'Subroutine FINITE calculates for finite source   $$$$$$$
4710 PRINT: INPUT"Size of source ";M
4720 IF M>1 THEN 4750
4730 PRINT "We need the NUMBER of units in the source!!!!!!"
4740 GOTO 4710
4750 SM=0
4760 FOR N=0 TO M
4770    F=M
4780    GOSUB 4610   '''''''''FACTORIAL
4790    NU#=FF#
4800    F=M-N
4810    GOSUB 4610   '''''''''FACTORIAL
4820    DE#=FF#
4830    E=(AR/NS)^N
4840    T(N)=NU#*E/DE#
4850    IF S=6 THEN GOSUB 5110   '''''''''ADJUST FOR MULT CHANNELS
4860    SM=SM+T(N)
4870 NEXT N
4880 P(0)=1/SM
4890 Q=0
4900 FOR N=1 TO M
4910    P(N)=T(N)*P(0)
4920    Q=Q+(N-1)*P(N)
4930 NEXT N
4940 PRINT "On the average,";Q;"units will be waiting"
4950 PRINT "to be served.  The average wait will be"
4960 PRINT Q/AR;U$+"S in the queue."
4970 PRINT
4980 RETURN
5000 'Subroutine FINTAB displays prob table for finite source  $$$$$$$$$$
5010 FOR N=0 TO M
5020    PRINT TAB(6) N; TAB(27) P(N)
```

Figure 10-5. (continued)

```
5030 NEXT N
5040 PRINT: RETURN
5100 'Subroutine FINMULT modifies FINITE for multiple channels  $$$$$$$$$
5110 IF N>CH THEN 5180
5120 'N <= the number of channels  <<<<<<<<<<<<
5130 F=N
5140 GOSUB 4610  '''''''''FACTORIAL
5150 CF#=FF#
5160 GOTO 5220
5170 'N > the number of channels  <<<<<<<<<<<<
5180 F=CH
5190 GOSUB 4610  '''''''''''FACTORIAL
5200 CF#=FF#*CH^(N-CH)
5210 'Modify single channel calculation  <<<<<<<<<<<<<
5220 T(N)=T(N)/CF#
5230 RETURN
```

just going to copy and use the program, you can save time by not copying the remarks. However, if you plan to make any changes to the program or if you want to follow it just to understand its operation, the remarks will help you locate certain sequences.

Because some versions of BASIC limit variable names to two characters, that limit is used in this program. If your computer accepts longer names, you can easily give the variables more descriptive identifications.

How the Program Works

This program is structured with a management section that includes lines 10 through 1020, and subroutines that do most of the detailed work. We explain the subroutines first because they are referred to often in explaining the management section.

Subroutines

Each subroutine begins with a remark that names and describes the subroutine. You can omit the name, as its only purpose is to aid in understanding the purpose of the subroutine. If you choose to delete the entire remark line, you do not have to change any part of the management section; all subroutine calls go to the first active line of the subroutine. The remark line that begins a subroutine ends with a series of $$$ to make it easy to locate the beginning of a subroutine.

Subroutine COMCALC, lines 4000–4060. This subroutine makes calculations that are used in several formulas in different parts of the program. Some of the results are modified by the management section after return from the subroutine.

Subroutine DISPCALC, lines 4100–4220. This subroutine prints the sentences that display the results. Numerical results are embedded in the sentences.

Subroutine ASKPROB, lines 4300–4470. This sequence asks if the operator wants a probability table. String variable PT$ in line 4310 causes either "probability table" or "more probabilities" to be printed — depending on where in the program this subroutine was called. Lines 4340–4350 give the operator another chance when the response is not acceptable.

When the operator chooses to see a table, lines 4380–4430 ask for the range to be displayed. The only check conducted here is that the upper end of the range is higher than the lower end. Then lines 4440–4460 print table headings and underline them.

Subroutine FACT!, lines 4600–4660. These lines calculate the factorial of variable *F*, calling the result FF#. The symbol # on the end makes the variable double precision, because factorials can be very large numbers.

Line 4610 initially sets FF# to one. Then line 4620 bypasses any further calculations if *F* is either zero or one, as the factorial of either is one.

Subroutine FINITE, lines 4700–4980. This is a long subroutine because the calculations are more involved when the source is finite. It begins by asking the size of the source. The lines 4750–4880 calculate P_0, performing that long repetition that we demonstrated earlier in the chapter with a source of three.

As subroutine FACT! calculates the factorial of variable *F*, line 4770 sets *F* equal to *M*; then this subroutine will calculate *M* factorial. On return from subroutine FACT!, FF# is the value of *M* factorial. Now line 4790 saves that value as NU#, because FACT! will be called again (by line 4810), which will change the value of FF#.

Line 4840 calculates one term for the summation. If the calculations are for a multiple channel situation, line 4850 calls subroutine FINMULT to adjust that term. Then the term is added to the summation in line 4860. Finally, P_0 is calculated in line 4880.

As with the manual calculations, the other quantities are not difficult (lines 4910 and 4920) once P_0 is known.

Lines 4940–4960 print the results that were just calculated, and control returns to the management section of the program.

Subroutine FINTAB, lines 5000–5040. This subroutine is called immediately after return from FINITE; it prints the probability table.

Subroutine FINMULT, lines 5100–5230. This subroutine is called by subroutine FINITE when there are multiple service channels. Line 5110 selects the appropriate formula, depending on how the number in the system compares with the number of service channels. The calculations in this subroutine are used to modify each term in the summation that finds P_0.

Management Program

Line 70 dimensions two arrays that are used in finite source calculations. Although this program does not approach the computer's memory limitations, we planned for a maximum source of 100 units. Sources larger than that are usually treated as infinite. However, you can change this line to dimension for larger sources.

Lines 80–250 display the main selection list (menu), accept the operator's selection, and check that the selection is within the range offered.

Lines 270–310 ask the operator for values to use in the calculations. Then line 320 directs the program to a series of operations, according to the operator's selection. As most of the operations are documented with remarks in the program, we will point out just a few items.

After subroutine ASKPROB asks if the operator wants a table of probabilities, line 380 returns the program to the menu if that is the operator's choice. If the operator chooses a table, lines 390–430 calculate and print it.

If the analysis involves multiple channels from an infinite source, line 470 asks "How many channels?" immediately following the other values requested in lines 270–310.

Check back to the formulas for multiple channels from infinite source and note that there are two terms in the denominator of P_0. Lines 490–540 calculate the first term (the summation) and lines 550–570 calculate the second. Line 580 then calculates P_0.

Lines 590–640 make the other calculations, and line 650 calls subroutine DISPCALC to display the results.

E\$ at the end of line 660 is a dummy variable required by the syntax of the INPUT statement. Whether the operator presses another key first or not, pressing the Enter key returns the program to the menu.

For constant service time, the calculations of subroutine COMCALC are modified by lines 700–730.

Each section of the management program returns to the menu when the operator is finished and presses Enter.

Testing the Computer Program

Typographic errors are easy to make, so it is a good idea to test the computer program. Try it with each of the examples in this chapter. Figure 10-6 shows the first screen that should appear when running the program.

First enter some negative numbers and numbers larger than six to verify that lines 230 and 240 reject them. Then select choice number one. The next two paragraphs help you verify the first example in this chapter.

The screen will ask for the basic unit of time — type **HOUR**. Enter six for the mean arrival rate, and eight for the number served per hour. Almost immediately after you enter the last value, the results will appear on the screen. Check them with the example.

Respond to the question about probability table with various letters and numbers, to test that line 4340 will reject anything except P or R. Then type P to see the table. Enter a lower limit of one and upper limit of five to see and check the same values as found in the example.

Figure 10-6. First screen that appears when the program is running.

INFINITE SOURCE

Negative exponential service time

1   Single channel
2   Multiple channels

Constant service time

3   Single channel

Poisson service time

4   Single channel

FINITE SOURCE

Negative exponential service time

5   Single channel
6   Multiple channels

Select by number

After verifying the table, enter *R* to return to the menu. Select two to verify the next part of the program, using the next example. Continue checking choices through number five of the menu. An example for choice six is given in the next section.

The program is organized into sections, with many remarks to document what various parts of the program do. If any part of the program gives answers different from the example, it should be easy to localize the trouble to a sequence of lines. You may then be able to figure out what is wrong, or it may be easier to simply compare your listing of those lines to the listing printed in the

Figure 10-7. Example using multiple channels.

Mean number served per HOUR 6
How many channels ? 2
Size of source ? 8
On the average, 4.614146 units will be waiting to be served. The average wait will be .9228291 HOURS in the queue.

Number of units	Probability of that number being in the system.
0	$1.240926E-03$
1	$8.272838E-03$
2	$2.412911E-02$
3	$6.032278E-02$
4	.1256725
5	.2094541
6	.2618176
7	.2181813
8	$9.090888E-02$

Press < Enter> for main selection list

book. After the program reproduces the results of the examples, you are ready to use the program for examining your own queuing situations.

Another Computer Example: How to Test Finite Source and Multiple Service Channels Formulas

Here is an example to test the last set of formulas — finite source and multiple service channels. Each of the eight assembly areas in our production department has a messenger that carries parts orders to the stock room and returns with the material. On the average, five messengers arrive at the stock room every hour. The orders are large, and it takes one of the two stock clerks twenty minutes to fill an average order.

First, let us prepare values to answer the computer's questions. Use number six from the menu, finite source and multiple channels. The basic unit of time is an hour, and the mean arrival rate is five per hour. Two stock clerks can each serve three messengers per hour, so the mean number served per hour is six. As the stock clerks work independently for one messenger at a time, there are two service channels. The size of the source is eight.

Figure 10-7 shows the results you should see on the screen.

Figure 10-8. Adapting a general procedure to a specific application

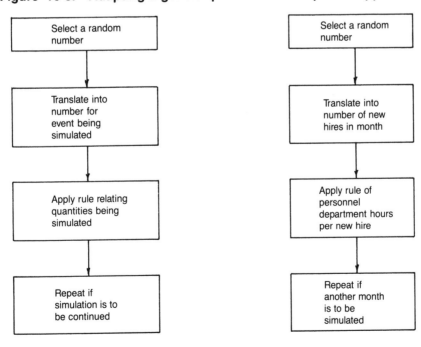

General   Specific

Table 10-7. Machine breakdown pattern without maintenance.

x	P(x)	Cumulative	Simulated by zero to . . .
0	.00674	.00674	7
1	.03369	.04043	40
2	.08422	.12465	125
3	.14037	.26503	265
4	.17547	.44049	440
5	.17547	.61596	616
6	.14622	.76218	762
7	.10445	.86663	867
8	.06528	.93191	932
9	.03627	.96817	968
10	.01813	.98631	986
11	.00822	.99455	995
12	.00343	.99798	998

SIMULATION: A POWERFUL TECHNIQUE

Simulation is a very powerful technique because it allows you to use any arrival and service time distributions, no matter how complex or unusual. Instead of being able to say that arrivals follow a Poisson or other distribution, you often have nothing but records of past arrivals. Those records are just as useful for simulation as a formula that describes arrivals and other conditions. The basic idea behind simulation is very simple — arrange for random numbers to trans-

Table 10-8. Machine breakdown pattern with maintenance.

x	P(x)	Cumulative	Simulated by zero to . . .
0	.00035	.00035	0
1	.00268	.00302	3
2	.01073	.01375	14
3	.02863	.04238	42
4	.05725	.09963	100
5	.09160	.19124	191
6	.12214	.31337	313
7	.13959	.45296	453
8	.13959	.59255	593
9	.12408	.71662	717
10	.09926	.81589	816
11	.07219	.88808	888
12	.04813	.93620	936

With or without maintenance, each breakdown costs $273.

Table 10-9. Manual simulation for ten years.

	Without maintenance				With maintenance			
Year	RN	Re-pairs	Cost	Cum cost	RN	Re-pairs	Cost	Cum cost
1	158	5	$1365	$ 1365	487	5	$2121	$ 2121
2	796	10	2730	4095	199	3	1575	3696
3	12	2	546	4641	36	1	1029	4725
4	205	6	1638	6279	118	2	1302	6027
5	901	12	3276	9555	13	1	1029	7056
6	525	8	2184	11739	83	2	1302	8358
7	73	4	1092	12831	172	3	1575	9933
8	616	9	2457	15288	122	2	1302	11235
9	433	7	1911	17199	77	2	1302	12537
10	321	7	1911	19110	290	4	1848	14385

late into a given distribution; then establish rules that parallel the actual operation of the system being simulated. A few short examples are given now; Chapter 15 goes into more detail about preparing and running simulations.

Figure 10-8 shows the simulation procedure on the left. On the right the figure shows how a personnel department adapts the procedure to create a realistic pattern of its workload. They start by knowing how often they process certain numbers of new hires. For example, records might show that ten percent of the time they process four new hires in a month, twenty percent of the time they process seven new hires, thirty-eight percent of the time they process twenty-five, and so on. To simulate this pattern they would draw numbers one through 100 randomly, and say that if the number is one through ten, they process four new hires. (The probability of drawing a number between one and ten is the same as the probability of having a month in which they process four new hires.) Drawing a number between eleven and thirty represents processing seven new hires and so forth.

The best way to explain simulation is with examples. Our first example is easy enough to do with pencil and paper; the next one includes a computer program.

Start a simulation by converting all distributions to cumulative distributions.

Figure 10-9. Distribution of arrival times.

Each asterisk represents .5

0.0 to	0.1	**
0.1 to	0.2	******
0.2 to	0.3	************
0.3 to	0.4	****************
0.4 to	0.5	********************
0.5 to	0.6	****************
0.6 to	0.7	************
0.7 to	0.8	******
0.8 to	0.9	****
0.9 to	1.0	**

That is, prepare a table such that each number is given a probability of that number or less.

Manual Example of Simulation

You have a production area with a large number of machines that have a breakdown pattern shown by the first two columns of Table 10-7. The third column is the cumulative sum needed for simulation; for example, across from $x = 2$ is the sum of probabilities of $x = 0$, $x = 1$, and $x = 2$. As x approaches 12, the cumulative probability approaches certainty (one). That is, it is almost a certainty that the number of machines broken down will be between zero and twelve.

Your production department has further determined that the breakdown distribution will be as shown in Table 10-8 if you institute a maintenance program that costs $756 a year. It is not important for this simulation, but these distributions happen to be Poisson with means of eight and five respectively.

The fourth column in Tables 10-7 and 10-8 shows the random integers that translate to the cumulative probability values. They were obtained by moving the decimal point three places to the right and then accumulating a sum while moving down the column. They mean, for example, that a random number of 158 will translate to five breakdowns without maintenance and three breakdowns with maintenance (in Table 10-8, 158 is larger than 100 and less than 191; 158 translates to five breakdowns without maintenance).

Do not interpret moving the decimal point three places as a general rule for obtaining a range of random numbers. It was convenient here, but you may find that another procedure is best in another situation.

Now you simply pick a random number between 0 and 998, use Table 10-7 to translate it to a number of breakdowns, and multiply it by $273 to compute the cost for a year without maintenance. Then, for a year with the maintenance program, pick a random number between 0 and 936, use Table 10-8 to translate it to a number of breakdowns, and add $756.

Repeat the process for each year to be simulated. Table 10-9 shows the simulation for ten years. This table was not contrived by deliberate, rather than ran-

Table 10-10. Cumulative distribution of arrival times.

Time since last arrival	Cumulative frequency
0.1 or less	1
0.2 or less	4
0.3 or less	10
0.4 or less	18
0.5 or less	28
0.6 or less	36
0.7 or less	42
0.8 or less	46
0.9 or less	48
1.0 or less	49

Figure 10-10. Manual and computer simulations follow the same flow.

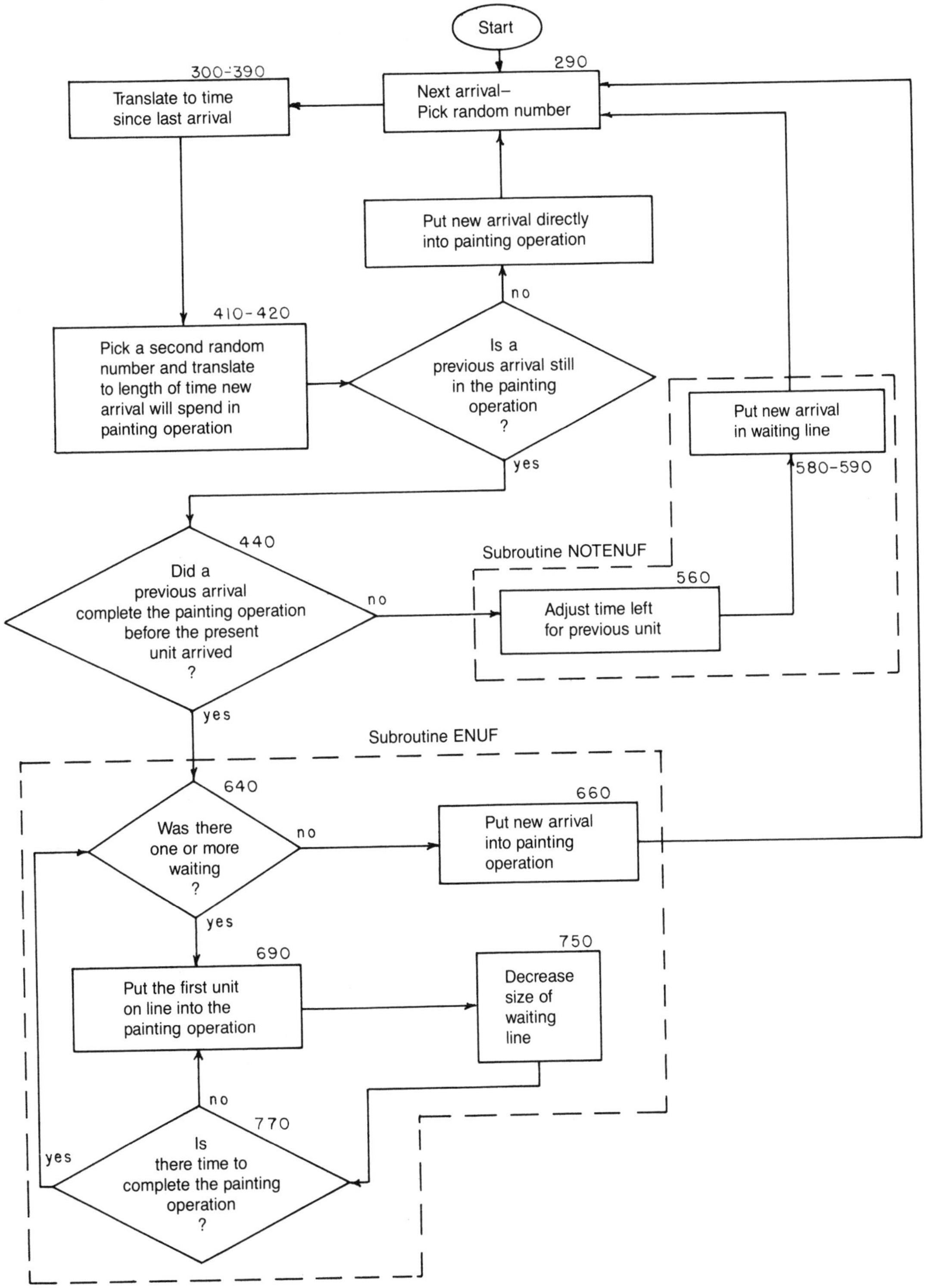

Figure 10-11. Computer program simulating queuing.

```
10 '******************************************************************
20 '** QSIM.BAS  Queuing simulation for Handbook of Manufacturing  **
30 '** and Production Management Formulas, Charts, and Tables       **
40 '******************************************************************
50 UI$="###"
60 U1$="#.##"
70 U2$="##.##"
80 DIM Q(500)
90 RANDOMIZE TIMER
100 CLS
110 PLACE=0
120 TIMELEFT=0
130 INPUT "Mean service time (.3 to 3) ";MST
140 IF MST>=.3 AND MST <=3 THEN 170
150 PRINT "Your selection must be between .3 and 3"
160 GOTO 130
170 INPUT "Shortest service time ";SST
180 IF SST>=0 AND SST<=MST THEN 220
190 PRINT "That can't be, since the mean"
200 PRINT "service time is";MST
210 GOTO 170
220 INPUT "How many arrivals are to be simulated ";N
230 PRINT TAB(2)"Unit";TAB(11)"Time since";TAB(25)"Service";TAB(37)"Time left fo
r";TAB(56)"Number"
240 PRINT "number";TAB(11)"last unit";TAB(27)"time";TAB(36)"unit in service";TAB
(56)"on line"
250 MST=MST*10
260 SST=SST*10
270 FOR ITEM=1 TO N
280 'Random number between 1 and 49  <<<<<<<<<<
290 RN=INT(RND*49)+1
300 IF RN<2 THEN TIMEIN=.1: GOTO 410
310 IF RN<5 THEN TIMEIN=.2: GOTO 410
320 IF RN<11 THEN TIMEIN=.3: GOTO 410
330 IF RN<19 THEN TIMEIN=.4: GOTO 410
340 IF RN<29 THEN TIMEIN=.5: GOTO 410
350 IF RN<37 THEN TIMEIN=.6: GOTO 410
360 IF RN<43 THEN TIMEIN=.7: GOTO 410
370 IF RN<47 THEN TIMEIN=.8: GOTO 410
380 IF RN<49 THEN TIMEIN=.9: GOTO 410
390 TIMEIN=1
400 'Find service time  <<<<<<<<<<<<<<<<
410 SRANGE=2*(MST-SST)+1
420 SERVTIME=(INT(RND*SRANGE)+SST)/10
430 TOGGLE=0
440 'Was there time to complete the unit in service?  <<<<<<<<<<<
450 IF TIMEIN>=TIMELEFT THEN GOSUB 630 ELSE GOSUB 570
460 'Print and get another arrival  <<<<<<<<<<<<<<<<
470 PRINT TAB(2) USING UI$; ITEM;
480 PRINT TAB(14) USING U1$; TIMEIN;
490 PRINT TAB(27) USING U1$; SERVTIME;
500 PRINT TAB(41) USING U2$; TIMELEFT;
510 PRINT TAB(57) USING UI$; PLACE
520 NEXT ITEM
530 INPUT "Press <Enter> to run again ",E$
540 GOTO 100
550 END
560 'Subroutine NOTENUF; not enough time to complete unit in serv  $$$$$$$
570 TIMELEFT=TIMELEFT-TIMEIN
580 'Put new arrival in queue  <<<<<<<<<<
```

Figure 10-11. (continued)

```
590 PLACE=PLACE+1
600 Q(PLACE)=SERVTIME
610 RETURN
620 'Subroutine ENUF; enough time to complete unit in service   $$$$$$$$
630 REMTIME=TIMEIN
640 'Is there a queue?   <<<<<<<<<<
650 IF PLACE>0 THEN 700
660 'No queue; put new arrival into service   <<<<<<<<<<
670 TIMELEFT=SERVTIME
680 GOTO 830
690 'There is a queue; put next unit into service   <<<<<<<<<<
700 REMTIME=REMTIME-TIMELEFT
710 TIMELEFT=Q(1)
720 FOR C=1 TO PLACE-1
730   Q(C)=Q(C+1)
740 NEXT C
750 Q(PLACE)=0
760 PLACE=PLACE-1
770 'Is there time to complete this one?   <<<<<<<<<<<<
780 IF REMTIME+.0001>=TIMELEFT THEN 650
790 'Not enough time; adjust TIMELEFT   <<<<<<<<<<<<
800 TIMELEFT=TIMELEFT-REMTIME
810 'Put new arrival on queue   <<<<<<<<<<<
820 GOSUB 590
830 RETURN
```

Table 10-11.   Waiting line builds up slowly when service time is about the same as arrival time.

Mean service time (.3 to 3)? .5
Shortest service time? .4
How many arrivals are to be simulated? 18

Unit number	Time since last unit	Service time	Time left for unit in service	Number on line
1	0.60	0.50	0.50	0
2	0.40	0.60	0.10	1
3	0.60	0.60	0.10	1
4	0.50	0.60	0.20	1
5	0.90	0.40	0.40	0
6	0.60	0.50	0.50	0
7	0.50	0.60	0.60	0
8	0.40	0.50	0.20	1
9	0.80	0.60	0.60	0
10	0.60	0.50	0.50	0
11	0.20	0.50	0.30	1
12	0.30	0.50	0.50	1
13	0.80	0.40	0.20	1
14	1.00	0.40	0.40	0
15	0.30	0.60	0.10	1
16	0.50	0.60	0.20	1
17	0.60	0.60	0.20	1
18	0.30	0.50	0.50	1

Press < Enter> to run again

Table 10-12.   Waiting lines begin to grow when service time exceeds arrival time.

Mean service time (.3 to 3)? .6
Shortest service time? .4
How many arrivals are to be simulated? 18

Unit number	Time since last unit	Service time	Time left for unit in service	Number on line
1	0.40	0.60	0.60	0
2	0.70	0.50	0.50	0
3	0.20	0.70	0.30	1
4	0.60	0.40	0.40	1
5	0.20	0.70	0.20	2
6	0.50	0.60	0.10	2
7	0.40	0.60	0.40	2
8	1.00	0.60	0.60	1
9	0.50	0.40	0.10	2
10	0.30	0.50	0.40	2
11	0.50	0.60	0.30	2
12	0.50	0.70	0.30	2
13	0.10	0.80	0.20	3
14	1.00	0.40	0.50	2
15	1.00	0.50	0.30	2
16	0.40	0.60	0.30	2
17	0.30	0.80	0.00	3
18	0.60	0.80	0.50	2

Press < Enter> to run again

dom, selection of numbers. However, mean values and annual maintenance costs were selected to make it likely that some years would favor one plan and some years the other.

Nothing more than pencil, paper, and a random number source are required for this simple simulation. There is a random number table in the appendix of this book. Most computers have a random number function that will give a string of pseudo-random numbers within a selected range. It is important to use a random number source, because humans are extremely poor at picking numbers out of the air without introducing a bias.

The longer you run a simulation, the more likely it is that it will represent a real life outcome. In our simulation of ten years, there were only six years in which the high cumulative cost did not switch sides. Therefore, many investigators would not feel that the simulation gave a strong enough indication of which plan to use. They would run some more simulations of ten years each to see if they all stabilized in the same way.

But even this short simulation of ten years becomes tedious and expensive when repeated several times. Then, too, you often have to simulate more than ten cycles — one year of a tool counter's daily operations requires over 250 cycles. Add to that the fact that most situations are much more complex than the one in this example, and you have a strong case for doing simulations by computer.

Computer Simulation

The method is simply to write down the steps you would follow if running the simulation manually; then tell the computer to do the same. The following situation demonstrates how a computer program devleops from analysis of the actions.

Tote boxes full of subassemblies from your assembly lines arrive at the paint booth at intervals described by Figure 10-9. At the two extremes, you can allow full drying time in the booth (resulting in longest service time) or you can keep the units just long enough to paint. If you dry them in the paint booth, boxes of subassemblies will arrive faster than you can handle them, causing an increase in work-in-process inventory costs. If you move them out of the booth as soon as they are painted, you will have to build a clean drying facility with filtered air. There are other choices between those extremes. You can allow vari-

Table 10-13. Still longer service time leads to still longer waits.

Mean service time (.3 to 3)? .8
Shortest service time? .6
How many arrivals are to be simulated? 18

Unit number	Time since last unit	Service time	Time left for unit in service	Number on line
1	0.70	0.70	0.70	0
2	0.70	0.60	0.60	0
3	0.60	0.80	0.80	0
4	0.80	0.60	0.60	0
5	0.20	1.00	0.40	1
6	0.30	0.90	0.10	2
7	0.50	0.90	0.60	2
8	0.70	1.00	0.80	2
9	0.80	0.90	0.90	2
10	0.40	1.00	0.50	3
11	0.30	0.70	0.20	4
12	0.60	0.90	0.60	4
13	0.90	0.60	0.60	4
14	0.50	0.80	0.10	5
15	0.30	0.90	0.80	5
16	0.30	0.60	0.50	6
17	0.60	0.80	0.60	6
18	0.60	0.90	0.90	6

Press < Enter> to run again

ous amounts of partial drying in the booth; each different time can be paired with a certain cost of special handling.

Because service time is under your control, you want to do this simulation several times, each with a different service time. Then you can see what waiting lines each causes, and be able to pick the arrangement promising lowest total costs.

Let us take service time as nearly a constant. It can be a mean value with a selected amount of random variation about the mean.

Figure 10-10 shows the flow of thought you would follow if doing a manual simulation. This should follow the real life flow of actions in the situation you want to simulate. Disregard for now the dashed lines and the numbers by the boxes; they refer to the computer program.

The diamond boxes are decision points, where the next action depends on conditions that exist when you reach the box. For example, when you arrive at the first diamond box, you have to know if the unit that was in service since the last arrival (if any) finished in the meantime.

Table 10-14. Units in line mean overtime pay at the end of the shift.

Mean service time (.3 to 3)? 1
Shortest service time? .8
How many arrivals are to be simulated? 18

Unit number	Time since last unit	Service time	Time left for unit in service	Number on line
1	0.80	1.20	1.20	0
2	0.70	0.80	0.50	1
3	0.60	0.80	0.70	1
4	0.60	1.20	0.10	2
5	0.60	0.80	0.30	2
6	0.60	0.90	0.90	2
7	0.70	1.10	0.20	3
8	0.20	0.90	0.00	4
9	0.20	1.00	0.60	4
10	0.40	0.80	0.20	5
11	0.10	1.10	0.10	6
12	0.50	1.00	0.50	6
13	0.50	0.90	1.10	6
14	0.20	0.80	0.90	7
15	0.50	1.10	0.40	8
16	0.40	0.90	0.00	9
17	0.70	0.90	0.20	9
18	0.60	1.10	0.60	9

Press < Enter> to run again

Computer Program

Let's first convert the arrival time distribution, Figure 10-9, into a cumulative distribution, as you did for the manual simulation. The results are shown in Table 10-10.

Now random numbers between 1 and 49 can translate into arrival times that will follow the pattern of Figure 10-9.

Service time is easier to work out because it is evenly distributed over an interval in this particular example. As it is the variable under your control, you will ask the operator for the mean value of service time. Then you will ask the operator for the shortest time a unit could be in service. The difference between those two figures will be one-half the range of service times. Let us say that the mean service time is 0.7 and the shortest is 0.6. Then one half the range is 0.1 and the whole range is 0.2. For each cycle of the simulation you will pick a random number netween 0 and 0.2 and add it to 0.6. The result will be your service time for that cycle (see lines 410 and 420 of the program).

Table 10-15. Choosing the same value for shortest and mean service times simulates constant service time.

Mean service time (.3 to 3)? .8
Shortest service time? .8
How many arrivals are to be simulated? 18

Unit number	Time since last unit	Service time	Time left for unit in service	Number on line
1	0.70	0.80	0.80	0
2	0.50	0.80	0.30	1
3	0.70	0.80	0.40	1
4	0.50	0.80	0.70	1
5	0.30	0.80	0.40	2
6	0.60	0.80	0.60	2
7	0.50	0.80	0.10	3
8	0.50	0.80	0.40	3
9	1.00	0.80	0.20	3
10	0.70	0.80	0.30	3
11	0.40	0.80	0.70	3
12	0.40	0.80	0.30	4
13	0.20	0.80	0.10	5
14	0.30	0.80	0.60	5
15	0.40	0.80	0.20	6
16	0.60	0.80	0.40	6
17	0.60	0.80	0.60	6
18	0.40	0.80	0.20	7

Press < Enter> to run again

Integers are easier to work with, but our example uses numbers such as 0.1, 0.2, and so on. Have the computer program multiply all these numbers by ten, then divide the final result by ten.

Those were the only special items in the computer program. All that is left now is to follow the logic of Figure 10-10 and write computer commands that will perform the actions in the boxes.

Figure 10-11 lists the entire program. It is structured with subroutines as indicated by dashed lines in Figure 10-10. Sequences of line numbers that correspond to each box are also shown.

Running Simulations

Tables 10-11 through 10-15 show five simulations of eighteen arrivals. As expected, longer service times result in longer waiting lines. Also note that the most frequent arrival time in Figure 10-9 is 0.5, and that waiting lines build up as service time exceeds that value.

Table 10-15 shows that this program allows you to simulate a truly constant service time by making the shortest time the same as the mean time.

Now you can quickly, easily, and inexpensively compare total costs of various arrangements. Your selection of the best policy can be justified.

Chapter 14 discusses simulation further. It also gives examples for other types of situations.

CHAPTER 11

Business Statistics: A Descriptive Approach to Collecting, Summarizing, and Arranging Data

This chapter looks at *descriptive* statistics—numbers that are merely collected, possibly grouped and combined, arranged, and presented. It also shows how to perform calculations on the data that are collected.

The section about collecting data discusses individual and grouped values, paying attention to discrete and continuous data, intervals, and their breakpoints, and includes a computer program for counting data values in intervals.

The next section on summarizing shows the three types of averaging techniques and how they work: arithmetic mean, median, and mode, as well as how to calculate weighted means. Also discussed are variance and standard deviation, two measures of data dispersion.

Next you will learn how to arrange the collected data for numbers in sequence, permutations (when placement matters), factorial, and combinations (when placement doesn't matter).

Computer programs and numerous examples are provided to help clarify the information included.

COLLECTING DATA: INDIVIDUAL AND GROUPED VALUES

A survey of hourly wages paid in production departments across the nation would record numbers such as $8.57, $12.16, and $5.35. If thousands of workers were surveyed you could expect to have hundreds of different wage rates. It would be difficult to get much meaning from such a long list, and you probably would not care for such details as how many earned $9.35, how many earned $9.36, and so forth.

With the survey results in hand, most people would follow these steps (even more likely—they would have their computer follow these steps):

1. scan the list to find the lowest and highest values
2. establish intervals of, say, $0.50

For example, if the lowest wage in the list was $3.43, you could set up groups of $3.00 to $3.49, $3.50 to 3.99, and so forth. Then, when you find the number earning in each group, you will have some facts that the mind is capable of assimilating.

Discrete data

The data as collected are called *individual values* and the last arrangement of data is called *grouped data*. Another important classification involves the possible closeness between data values. For example, dollars and cents are called *discrete data* because there is a difference of at least $.01 between different values. When setting up groups, it is safe to say that one group ended with $3.49 and the next began with $3.50, because there is no dollars and cents figure between them.

Continuous data

On the other hand, if the data values are average pieces produced by workers each day, the results can be carried to many decimal places and there is no minimum difference between data values. One day there could be 1819 pieces by 107 workers for an average of 17.000000; the next day there might be 1820 pieces by the same workers for an average of 17.009346 (to six places). But that does not even mean that 0.009346 is the minimum increment and no average could fall between those two. Suppose one day there were 108 workers present and they made 1837 units—the average would be 17.009259 (to six places). When the data is such that another value can always fit between any two values, it is called *continuous data*.

Of course, an analyst can make the data into discrete data by limiting the number of decimal places, as is effectively done with dollars and cents. But if continuous data is used, it must be handled differently.

Intervals

For one thing, the only way to combine continuous data is in intervals. Returning to the factory wage survey, notice that individual data values could have been combined to find, say, 1,105 people earning $7.69, 984 people earning $7.68. But with continuous data it is unlikely that any data value would appear more than once.

Breakpoints for intervals are easy with discrete data because you can specify the end of one interval and the beginning of the next, and know that no data value will drop in between them. You cannot do that with continuous data, so a convention has been established: you specify the same number as the top of one interval and the bottom of the next, but understand that the interval actually ends at the top limit; the next interval actually begins with "larger than" the specified value. For example, average worker output could be grouped 15 to 17, 17 to 19, 19 to 21, and so on. Then 16.9, 16.999999, and 17.000000 would all be counted in the first group and 17.000001 would be counted in the second group. But you could not say that the first group ends with 17.000000 and the second begins with 17.000001 because there might be a value of 17.0000001 that would fall between 17.000000 and 17.000001.

Example 11.1

A firm uses a dumbwaiter to carry material between the floors of the production department. Blocks are provided on each floor so that the door can be held open during loading and unloading. There is some concern about the lengths of time the dumbwaiter is out of service, so a team has been assigned to investigate the situation.

As part of the investigation, the team has checked the number of times the dumbwaiter was out of service in each of forty days. They also timed each tie-up and calculated the average length of time. The results are shown in Table 11-1.

Then they divided the average times into intervals of 0.2 minutes and counted the number of averages that fell into each interval. The results are shown in Table 11-2.

With the data counted in groups the team can see how many tie-ups take various lengths of time. It would be difficult to get any useful information from the raw data as it was collected.

COMPUTER PROGRAM FOR COUNTING DATA VALUES IN INTERVALS

The computer program listed in Figure 11-1 speeds up the task of counting the number of data values in each interval. It therefore makes it easy to experiment with interval sizes and have the computer take new counts.

Menu choices

The menu divides the choices into two groups: four choices connected with input and output of data, and three choices on handling data in memory. The subroutine that displays raw data on the screen arranges two sets of columns so as to avoid wasting a large part of the screen area. It also prints one full screen at a time so that results do not scroll off without giving you a chance to study them.

Choice number four saves the data in memory. If this choice were not in the program, it would be necessary to reenter the same set of data every time you turn on the computer or run the program. (Typing **RUN** to start the program erases all data in memory.) When entering data, there is always a chance of error, but loading data from a disk loads it exactly as it was saved.

This program saves, for each day, only the number of tie-ups and the total time tied up. Average times of tie-ups and numbers in intervals are not saved because they would take extra space on the disk. It is not necessary to save them because the computer easily calculates them from the numbers that are saved.

Example 11.2

Study the data of the preceding example with interval widths of 0.5 minutes.

Run the program and then select number one (enter all new data from keyboard) from the menu. It is a good practice to save data as soon as it is entered (choice number three), so that it can be recalled and used another time. However, it is not necessary to reload the data each time you change something such as interval size.

After you enter the last set of data values (you tell the computer there is no more data by typing **END** when it asks for the next set), you should select number five to check the data. If it is all correct, select number six. When the screen asks for interval size, type .5 and press **ENTER**. The computer will indicate when it has finished grouping the data; press **ENTER** to display the menu again, and then select number seven.

Table 11-3 shows the screen with data regrouped into intervals of 0.5 minutes each.

Table 11-1. Dumbwaiter tie-ups for 40 days.

Day	Tie-ups	Total time	Mean time	Day	Tie-ups	Total time	Mean time
1	15	252.19	16.812670	21	19	316.35	16.650000
2	3	49.17	16.390000	22	16	274.00	17.125000
3	12	200.31	16.692500	23	17	278.89	16.405300
4	12	201.27	16.772500	24	10	168.31	16.831000
5	15	248.99	16.599330	25	13	222.57	17.120770
6	6	106.05	17.675000	26	16	270.65	16.915630
7	6	97.57	16.261670	27	16	266.93	16.683130
8	6	91.60	15.266670	28	4	71.60	17.900000
9	12	194.80	16.233330	29	17	296.49	17.440590
10	13	208.41	16.031540	30	4	66.71	16.677500
11	14	234.57	16.755000	31	17	283.56	16.680000
12	14	232.89	16.635000	32	18	302.28	16.793330
13	4	66.70	16.675000	33	16	262.51	16.406880
14	8	135.45	16.931250	34	12	196.19	16.349170
15	7	115.47	16.495710	35	7	120.70	17.242860
16	20	337.63	16.881500	36	11	189.76	17.250910
17	11	182.79	16.617270	37	12	202.58	16.881670
18	15	251.36	16.757330	38	12	205.69	17.140830
19	17	281.94	16.584710	39	17	289.00	17.000000
20	6	97.90	16.316660	40	5	89.02	17.804000

SUMMARIZING DATA AND MAKING COMPARISONS

Three Types of Averages and How They Work

One frequent reason for collecting data is to make comparisons with other sets of data. But how should the comparison be made? Compare the lowest val-

Table 11-2. Data grouped into intervals of 0.2.

15.0	to 15.2	0
15.2	to 15.4	1
15.4	to 15.6	0
15.6	to 15.8	0
15.8	to 16.0	0
16.0	to 16.2	1
16.2	to 16.4	5
16.4	to 16.6	5
16.6	to 16.8	12
16.8	to 17.0	7
17.0	to 17.2	3
17.2	to 17.4	2
17.4	to 17.6	1
17.6	to 17.8	1
17.8	to 18.0	2

Figure 11-1. Computer program grouping continuous data into intervals

```
10 '********************************************************
20 '** CONDAT.BAS  Does calculations with continuous data  **
30 '** for Statistics chapter of Handbook of Manufacturing **
40 '** and Production Management Formulas, Charts, and      **
50 '** Tables                                               **
60 '********************************************************
70 SCREEN 6
80 DIM NT(200), TT(200), AT(200), SUMG(200)
90 UI$="###": U2$="###.##": UA$="##.######"
100 CLS
110 GOSUB 170              'Display menu
120 CLS
130 ON SELECT GOSUB 320,330,440,550,630,960,1280
140 GOTO 100
150 END
160 'Sbr to display menu  <<<<<<<<<<<<<<<
170 PRINT
180 PRINT TAB(20)"Data input/output"
190 PRINT TAB(8)" 1 Enter all new data from keyboard"
200 PRINT TAB(8)" 2 Add keyboard entries to data in memory"
210 PRINT TAB(8)" 3 Retrieve data from disk"
220 PRINT TAB(8)" 4 Save data in disk"
230 PRINT TAB(20)"Operations on data in memory"
240 PRINT TAB(8)" 5 Display raw data on screen"
250 PRINT TAB(8)" 6 Combine data into groups"
260 PRINT TAB(8)" 7 Display grouped data
270 INPUT "Select by number ",SELECT
280 IF SELECT>0 AND SELECT<8 THEN RETURN
290 PRINT "Select only 1 through 9"
300 PRINT :GOTO 270
310 '(1) and (2) Sbr to enter data from keyboard  <<<<<<<<<<<<<<
320 N=0                    'Enter here for selection #1, next line for 2
330 PRINT "Enter number of tie-ups for day #";N+1;"... or END";
340 INPUT " ",T$
350 IF T$="END" THEN RETURN
360 N=N+1
370 NT(N)=VAL(T$)
380 PRINT "Enter total time of tie-ups for day #";N;
390 INPUT " ",TT(N)
400 AT(N)=TT(N)/NT(N)
410 PRINT
420 GOTO 330
430 '(3) Sbr to load data from disk  333333333333333333
440 INPUT "Name of file on disk ",N$
450 OPEN "I",#1,N$
460 N=0
470 IF EOF(1) THEN 520
480 N=N+1
490 INPUT #1, NT(N),TT(N)
500 AT(N)=TT(N)/NT(N)
510 GOTO 470
520 CLOSE
530 RETURN
540 '(4) Sbr to save data on disk  4444444444444444
550 INPUT "Name to be used for saving data ",N$
560 OPEN "O",#2,N$
570 FOR C=1 TO N
580    PRINT #2,NT(C),TT(C)
590 NEXT C
600 CLOSE
```

Figure 11-1. (continued)

```
610 RETURN
620 '(5) Sbr to display raw data on screen   5555555555555
630 NBREM=N
640 FIRST=1
650 MORE=1
660 IF NBREM>40 THEN 700
670 NBRNOW=NBREM
680 MORE=0
690 GOTO 710
700 NBRNOW=40
710 ROWS=INT((NBRNOW+1)/2)
720 PRINT TAB(15)"Total";TAB(25)"Average";TAB(55)"Total";TAB(65)"Average"
730 PRINT "Day  Tie-ups  time        time";
740 PRINT TAB(40)"Day  Tie-ups    time         time"
750 FOR C=FIRST TO FIRST+ROWS-1
760    PRINT TAB(1) USING UI$;C;
770    PRINT TAB(9) USING UI$;NT(C);
780    PRINT TAB(15) USING U2$;TT(C);
790    PRINT TAB(24) USING UA$;AT(C);
800    CR=C+ROWS
810    IF MORE=0 AND CR>N THEN 860
820    PRINT TAB(40) USING UI$;CR;
830    PRINT TAB(48) USING UI$;NT(CR);
840    PRINT TAB(54) USING U2$;TT(CR);
850    PRINT TAB(63) USING UA$;AT(CR);
860 NEXT C
870 PRINT
880 IF MORE=0 THEN 930
890 NBREM=NBREM-40
900 FIRST=FIRST+40
910 INPUT "Press ENTER for next screen ",E$
920 GOTO 660
930 INPUT "Press ENTER for main menu ",E$
940 RETURN
950 '(6) Sbr to combine data into groups  6666666666666666
960 INPUT "Enter low end of first group ",L
970 INPUT "Enter size of intervals ",I
980 BIG=AT(1)
990 FOR C=2 TO N
1000    IF AT(C)>BIG THEN BIG=AT(C)
1010 NEXT C
1020 TOP=L+I
1030 NG=1
1040 PRINT "Groups are:";
1050 PRINT TOP-I;"-";TOP;"   ";
1060 IF TOP>=BIG THEN 1100
1070 TOP=TOP+I
1080 NG=NG+1
1090 GOTO 1050
1100 PRINT
1110 FOR C=1 TO NG
1120    SUMG(C)=0
1130 NEXT C
1140 C=1
1150 TOP=L+I: GC=1
1160 IF AT(C)>TOP THEN 1210
1170 SUMG(GC)=SUMG(GC)+1
1180 IF C>=N THEN 1230
1190 C=C+1
1200 GOTO 1150
```

Figure 11-1. (continued)

```
1210 TOP=TOP+I :GC=GC+1
1220 GOTO 1160
1230 COMBINED=1
1240 PRINT :PRINT "Data grouping complete"
1250 INPUT "Press ENTER to return to menu ",E$
1260 RETURN
1270 '(7) Sbr to display grouped data on screen   77777777777777
1280 IF COMBINED=1 THEN 1320
1290 PRINT "Data in memory has not been grouped"
1300 INPUT "Press ENTER to return to menu ",E$
1310 GOTO 1410
1320 TOP=L
1330 FOR C=1 TO NG
1340    TOP=TOP+I
1350    PRINT USING UA$;TOP-I;
1360    PRINT " to ";
1370 PRINT USING UA$;TOP;
1380 PRINT TAB(27) USING UI$;SUMG(C)
1390 NEXT C
1400 INPUT "Press ENTER to return to menu ",E$
1410 RETURN
```

ues of each set, then the second values, and so forth? What if one set has 100 data values and the other has 537? The answer is that you need a number that summarizes each set of data and then you can compare the summaries.

Arithmetic mean

The most frequently used summary is the arithmetic mean, commonly called the average. The longer expression is used here because there are several kinds of averages.

$$m = \frac{v_1 + v_2 + v_3 + \cdots + v_n}{n}$$

where v_1, v_2, ... are the first, second, and so forth data values and n is the total number of data values.

Example 11.3

The number of production workers absent each day of the last two weeks was 3, 2, 3, 3, 5, 3, 1, 3, 2, and 5. What was the mean number absent?

$$m = \frac{3 + 2 + 3 + 3 + 5 + 3 + 1 + 3 + 2 + 5}{10}$$

$$= 3$$

Table 11-3. Computer program makes it easy to investigate many interval sizes.

15.0	to 15.5	1
15.5	to 16.0	0
16.0	to 16.5	9
16.5	to 17.0	21
17.0	to 17.5	6
17.5	to 18.0	3

Median

Another type of average requires that the data be arranged in ascending or descending order. The median is then the value that is in the middle. When the number of data values is odd, there is actually a middle value, and it is the median. An even number of data values does not have a middle value, and the median is considered to be between the two middle values — usually it is then taken as the arithmetic mean of the two. The arithmetic mean of two values is halfway between them.

Note that the median is not defined as being the midpoint of all data values — halfway between the highest and the lowest values. It is the midpoint in the sense that the number of data points above it equals the number below it.

Example 11.4

The number of employee cars in the parking lot on ten successive working days was 1,232, 1,401, 1,344, 1,311, 1,198, 1,240, 1,388, 1,392, 1,307, and 1,205. What is the median daily number?

First, arrange the data in numerical order. Then, as there is an even number of data values, take the median as the arithmetic mean of the two in the middle. The values in numeric order are 1,198, 1,205, 1,232, 1,240, 1,307, 1,311, 1,344, 1,388, 1,392, and 1,401. The fifth and sixth values are in the middle, so the median is $(1,307 + 1,311)/2 = 1,309$.

Notice that data values other than the middle ones do not affect the value of the median. Suppose that the smallest number had been 98 instead of 1,198 — it would not change the median. In fact, as long as the number of values below 1,307 is the same as the number of values above 1,311, the median will be 1,309.

Mode

The third measure that statisticians recognize as an average is the mode, which is the data value that appears most often.

Example 11.5

A toy manufacturer wants to make a special factory run of sets of plastic pieces that plug together to build various objects. Recent sales of their sets have been as shown in Table 11-4.

The only statistic that interests the firm is the fact that the 100 piece set was wanted by more customers than any other size. It is not helpful to know that the mean number of pieces in sets bought is 127.13. Neither is it helpful to know that the median size set is 125.

When the data values form two peaks, even if they are of different sizes, the distribution is said to be *bimodal*. For example, if the production department runs a class in fundamentals of electronics, it can expect the grades to be bimodal. The explanation here is that the grades of any group will usually have a mode, and the production department probably has two groups — those with some electronics experience and those with none.

Table 11-4. Recent sales of plastic toy.

Set size	Number sold
50 piece	312
75 piece	147
100 piece	868
125 piece	503
150 piece	594
175 piece	231
200 piece	435

Calculating the Weighted Mean

When figuring the arithmetic mean, equal consideration was given to all values. However, there are many situations in which some data points are more significant than others so they are given different *weights*. The arithmetic mean so calculated is the weighted mean. The formula is

$$m_w = \frac{w_1 v_1 + w_2 v_2 + \cdots + w_n v_n}{w_1 + w_2 + \cdots w_n}$$

where w_1, w_2, and so on are the weights assigned to values 1, 2, and the rest.

Example 11.6

Table 11-5 shows prices of some of the items often drawn from the tool counter.

Table 11-5. Items drawn at tool counter.

Item	Price	Quantity drawn
A	$0.75	48
B	9.64	11
C	2.39	72
D	36.40	3

Because the quantities used of some items are so much larger than quantities of others, a mean found by simply adding each price and dividing by four would not be helpful. A more meaningful summary figure can be calculated by weighting each price with the quantity used.

$$m_w = \frac{48(0.75) + 11(9.64) + 72(2.39) + 3(36.40)}{48 + 11 + 72 + 3}$$

$$= \$3.16$$

Measures of Data Dispersion, Discrete Data

Variance

In seeking numbers to summarize the data, we can use one of the averages to describe where most of the action is. But the average does not adequately summarize the data. Two distributions with the same mean would look quite different if all the data values of one are clustered near the mean and the other's are scattered widely. Variance is a measure of dispersion. Specifying one of the averages and the variance gives a good summary of the data.

The formula for variance is

$$var = \frac{(x_1 - \bar{x})^2 + (x_2 - \bar{x})^2 + \cdots (x_n - \bar{x})^2}{n}$$

where $\bar{x}$ is the arithmetic mean.

Example 11.7

Two machines are adjusted to produce pieces with the same dimension. As shown near the bottom of Table 11-6, these settings result in the same mean value for each machine. The "Size" columns in the table show the measurements of twenty pieces from each machine, and the "Diff" columns show the difference of each measurement from the mean of that machine.

The last column under each machine is the difference squared, for use in the formula. The sum of these squares, shown at the bottom of the column, divided by twenty (the number of data values) is the variance; it is displayed at the end of Table 11-6. Machine 2, which has more dispersion in the measurements of pieces it produces, has a larger variance. Therefore it can be expected that Machine 2 will produce more rejects, even if the mean is the same for pieces from either machine.

Standard Deviation

Standard deviation is the square root of variance and is used much more than variance as a measure of dispersion. In the preceding example, Machine 1 has a standard deviation of 0.00118 and Machine 2 has a standard deviation of .02442.

In a normal distribution it can be expected that about sixty-eight percent of all items fall within one standard deviation of center, ninety-six percent within two standard deviations, and over ninety-nine percent within three standard deviations.

An important feature of standard deviation is that it can be used to compare different sets of numbers, even if they do not have the same number of data values or the same mean.

Example 11.8

Machines 1 and 2 are now producing different items. As shown in Table 11-7, the inspector has measurements from fifteen pieces from Machine 1 and

Table 11-6. Machines produce same mean but different spreads.

	Machine 1				Machine 2		
Piece	Size	Diff	Diff sqd	Piece	Size	Diff	Diff sqd
1	14.507	0.007	0.000051	1	14.501	0.001	0.000001
2	14.492	−0.008	0.000062	2	14.508	0.008	0.000066
3	14.498	−0.002	0.000003	3	14.467	−0.033	0.001079
4	14.506	0.006	0.000038	4	14.477	−0.023	0.000522
5	14.518	0.018	0.000329	5	14.463	−0.037	0.001358
6	14.491	−0.009	0.000078	6	14.472	−0.028	0.000776
7	14.505	0.005	0.000027	7	14.514	0.014	0.000200
8	14.494	−0.006	0.000034	8	14.496	−0.004	0.000015
9	14.516	0.016	0.000261	9	14.520	0.020	0.000406
10	14.511	0.011	0.000124	10	14.527	0.027	0.000737
11	14.500	0.000	0.000000	11	14.506	0.006	0.000038
12	14.513	0.013	0.000173	12	14.473	−0.027	0.000721
13	14.488	−0.012	0.000140	13	14.526	0.026	0.000684
14	14.497	−0.003	0.000008	14	14.469	−0.031	0.000952
15	14.499	−0.001	0.000001	15	14.533	0.033	0.001099
16	14.517	0.017	0.000294	16	14.484	−0.016	0.000251
17	14.481	−0.019	0.000355	17	14.483	−0.017	0.000284
18	14.479	−0.021	0.000435	18	14.512	0.012	0.000148
19	14.481	−0.019	0.000355	19	14.519	0.019	0.000367
20	14.504	0.004	0.000017	20	14.547	0.047	0.002223

MEAN = 14.49985 Sum = 2.786513E-03 MEAN = 14.49985 Sum = 1.192658E-02
Variance = 1.393256E-04 Variance = 5.96329E-04

ten pieces from Machine 2. A piece from Machine 1 measures 14.507 and a piece from Machine 2 measures 30.0245 inches. Which piece should cause the inspector more concern?

The difference between the first piece and the mean of Machine 1 is 14.507 − 14.496, or .011. This difference is a little over one standard deviation (which is shown in the last line of Table 11-7 as .0107).

The difference between the second piece and the mean of Machine 2 is 30.0245 − 29.9995, or 0.0250. This difference is a little less than one standard deviation (.0269). Therefore, the first piece is further from normal because it is further in terms of standard deviation.

Measures of Dispersion, Grouped Data

Once data values have been grouped into intervals they lose their individual values. Every data point in an interval is assumed to have the same value, usually the midpont of the interval. Mean and standard deviation are then calculated accordingly.

Example 11.9

Find the mean and standard deviation of the data in Table 11-3.

The midpoints of the intervals are 15.25, 15.75, 16.25, 16.75, 17.25, and

Table 11-7. Standard deviation allows comparison of sets of data with different centers and sizes.

	Machine 1				Machine 2		
Piece	Size	Diff	Diff sqd	Piece	Size	Diff	Diff sqd
1	14.482	−0.014	0.000204	1	29.976	−0.024	0.000552
2	14.487	−0.009	0.000086	2	29.995	−0.004	0.000020
3	14.517	0.021	0.000430	3	29.956	−0.044	0.001892
4	14.490	−0.006	0.000039	4	29.991	−0.009	0.000072
5	14.496	−0.000	0.000000	5	30.009	0.010	0.000090
6	14.506	0.010	0.000095	6	30.016	0.017	0.000272
7	14.494	−0.002	0.000005	7	30.047	0.048	0.002256
8	14.515	0.019	0.000351	8	29.971	−0.028	0.000812
9	14.499	0.003	0.000007	9	30.035	0.035	0.001260
10	14.501	0.005	0.000022	10	29.999	−0.000	0.000000
11	14.487	−0.009	0.000086				
12	14.503	0.007	0.000045				
13	14.479	−0.017	0.000298				
14	14.489	−0.007	0.000053				
15	14.499	0.003	0.000007				

MEAN = 14.49627 Sum = 1.728915E-03   MEAN = 29.9995 Sum = 7.228642E-03
Variance = 1.15261E-04   Variance = 7.228642E-04
Standard deviation = 1.073597E-02   Standard deviation = 2.688613E-02

17.75. There is one data point with a value of 15.25; none in the second interval; nine in the third interval, all with the same value of 16.25; and so forth. First calculate the mean value so that we can find each interval's difference from the mean.

$$m = \frac{1(15.25) + 0(15.75) + 9(16.25) + 21(16.75) + 6(17.25) + 3(17.75)}{40}$$

$$= 16.75$$

Now subtract the mean value from each of the center values of the intervals. Then square the differences (results of the subtractions), and add the squares. Table 11-8 shows the calculations.

Table 11-8. Calculations for finding standard deviation of grouped data.

Interval	Center	Number	Difference	Diff Squared
15.0-15.5	15.25	1	−1.5	2.25
15.5-16.0	15.75	0	−1.0	1.00
16.0-16.5	16.25	9	−0.5	0.25
16.5-17.0	16.75	21	0	0
17.0-17.5	17.25	6	0.5	0.25
17.5-18.0	17.75	3	1.0	1.0

Now find the sum of squared differences by multiplying each squared difference by the number of times it occurs, and adding the products. Then variance is that sum divided by the total number of items. This procedure follows the details given previously, under the the section on *Variance*.

$$var = \frac{1(2.25) + 0(1.00) + 9(0.25) + 21(0) + 6(0.25) + 3(1.0)}{40}$$

$$= 0.225$$

and standard deviation is the square root of variance, or 0.474.

HOW TO CALCULATE THE NUMBER OF POSSIBLE ARRANGEMENTS FOR THE COLLECTED DATA

This section looks at the number of ways that a set of items can be arranged, first when placement is specified and then when it is not.

Figuring Events in Sequence

If an end result is due to a series of actions or events, then the number of end results there could be is the product of the numbers of ways each event can be done. For example, if the top of a container can be painted with any one of four colors, the sides with any of three colors, and the bottom with any one of six colors, then the number of different containers possible is 4(3)6, or 72.

Example 11.10

Promotional samples are to be prepared with four classes of your products. You can include five different types of class A products, three different types of class B, two different types of class C, and four different types of class D. How many different sample sets can you make up?

The number possible is the product of all the separate possibilities: 5(3)2(4), or 120 different sample sets.

Finding factorial

Calculating factorial is a necessary part of calculating possibilities and probabilities. Factorial for a given number is defined as the product of all integers from one through the given number. For example, four factorial is 1 times 2 times 3 times 4, or 24. Zero factorial is defined as equal to one.

The symbol for factorial is an exclamation mark. Six factorial is written 6!.

Permutations: Determining the Number of Arrangements When Positions Are Specified

A permutation is an arrangement in which the position of each item is specified. For example, suppose five people, Green, Smith, Jones, Doe, and Fields, complete a management training program and are available for assignment. You have three group leader positions to fill. If you pick Green for Group A, Smith for Group B, and Jones for Group C, that is one permutation of those three individuals. Switching Green to Group B and Smith to Group A creates another permutation. Statisticians say that Green, Smith, Jones is one permutation and Smith, Green, Jones is another. Had you picked the same three people but not specified how you would assign them, then Green, Smith, Jones would be the same as Smith, Green, Jones and you would not be looking at permutations.

Another way to look at this example is — if top management asked who you picked for Group A, who for Group B, and who for Group C, there are sixty answers you could give. But if they asked simply for the three people you picked, there are ten answers you could give (as you will see in the next section).

If there are n different items to choose from and you are to choose r of them, then the number of permutations possible is written $_nP_r$.

$$_nP_r = \frac{n!}{(n - r)!}$$

It is often referred to as "the number of permutations of n things taken r at a time."

Example 11.11

The public relations department asks you for six of your latest products. They plan to place four of them in a showcase. In how many different ways can they arrange the showcase?

This is another way of asking, "How many permutations are there of six things taken four at a time?"

$$_6P_4 = \frac{6!}{(6 - 4)!}$$

$$= \frac{720}{2}$$

$$= 360$$

Combinations: When Placement Doesn't Matter

A combination is a selection without regard to the order in which they are placed. Returning to the container with different colors for the top, sides, and bottom, suppose that the top is green, the sides are yellow, and the bottom is orange. If you interchange the top and sides — make the top yellow and the sides green — you have a new permutation but not a new combination. Using green, yellow, and orange without specifying which part of the container will have which color is a combination.

For any given number to choose from, you can make fewer combinations than you can permutations, so it is reasonable that the formula for combinations should have a larger denominator than the formula for permutations. The possible number of combinations of n things taken r at a time is written $_nC_r$.

$$_nC_r = \frac{n!}{r!\,(n-r)!}$$

Example 11.12

There are six items to go into an assembly, and three of them are to be placed in front of one assembler. How many permutations and how many combinations can be placed there?

The first question is, "How many permutations can we make of six things taken three at a time?" Numbers can be substituted directly into the formula

$$_6P_3 = \frac{6!}{(6-3)!}$$

$$= 120 \text{ permutations}$$

If you are not concerned with, for example, whether this assembler places the items in the order nuts-bolts-washers or washers-nut-bolts, then you are counting combinations. The number of combinations possible is

$$_6C_3 = \frac{6!}{3!\,(6-3)!}$$

$$= 20$$

Table 11-9 gives the number of combinations possible for up to fourteen items. It is a duplication of Table 6-3 in Chapter 6.

Table 11-9. Table of combinations

X N	0	1	2	3	4	5	6	7	8	9	10	11	12	13	14
1	1	1													
2	1	2	1												
3	1	3	3	1											
4	1	4	6	4	1										
5	1	5	10	10	5	1									
6	1	6	15	20	15	6	1								
7	1	7	21	35	35	21	7	1							
8	1	8	28	56	70	56	28	8	1						
9	1	9	36	84	126	126	84	36	9	1					
10	1	10	45	120	210	252	210	120	45	10	1				
11	1	11	55	165	330	462	462	330	165	55	11	1			
12	1	12	66	220	495	792	924	792	495	220	66	12	1		
13	1	13	78	286	715	1287	1716	1716	1287	715	286	78	13	1	
14	1	14	91	364	1001	2002	3003	3432	3003	2002	1001	364	91	14	1

COMPUTER PROGRAM FOR CALCULATING PERMUTATIONS AND COMBINATIONS

Figure 11-2 lists a program that calculates permutations and combinations. The results can be used in various applications.

Example 11.13

Four supervisors from the twelve in the production department will be selected to be on a new products committee. How many groups of four can be made? If a committee leader, assistant, liaison, and coordinator will be appointed by name, how many ways can the group be formed?

As the first question allows each of the four to take any position on the committee, it is asking for combinations. The second question is asking for permutations. Enter twelve when the computer program asks for total number, and four when it asks for the number to be selected. It will then show that there are

Figure 11-2. Computer program calculating permutations and combinations.

```
10 '****************************************************************
20 '** PERCOM.BAS  Calculates permutations and combinations   **
30 '** for Statistics chapter of Handbook of Manufacturing    **
40 '** and Production Management Formulas, Charts, and         **
50 '** Tables                                                  **
60 '****************************************************************
200 CLS
210 INPUT "How many items will be chosen? ",R
220 INPUT "From a pool of how many? ",N
230 IF N>=R THEN 280
240 PRINT
250 PRINT "You cannot choose";R;"items from a pool of";N
260 PRINT
270 GOTO 210
280 'Perform calculations   <<<<<<<<<<<<<<<
290 F=N
300 GOSUB 1010              'Calculate n!
310 NF=FF
320 F=R
330 GOSUB 1010              'Calculate r!
340 RF=FF
350 F=N-R
360 GOSUB 1010              'Calculate (n-r)!
370 P=NF/FF
380 C=P/RF
390 CLS
400 PRINT "There are";C;"combinations and";P;"permutations"
410 PRINT "of";N;"things taken";R;"at a time."
420 PRINT
430 INPUT "Press ENTER to make another calculation ",E$
440 GOTO 200
450 END
1000 'Subroutine to calculate factorials  $$$$$$$$$$$$$$$
1010 FF=1
1020 IF F<2 THEN 1060
1030 FOR C=2 TO F
1040    FF=FF*C
1050 NEXT C
1060 RETURN
```

495 groups that can be formed, and 1980 ways that four supervisors can be named to positions on the committee.

PROBABILITY: FIGURING YOUR CHANCES OF AN EVENT ACTUALLY HAPPENING

Probability is a number between zero and one, with zero meaning that the event could not possibly happen and one meaning that its happening is as certain as weeds in a garden.

Basic Ratio of Probability

Basically, probability is calculated by dividing the number of ways the given event could occur by the total number of outcomes that could occur.

Example 11.14

An assembler is given a bin of nuts, a bin of bolts, and a bin of washers to be placed in the most convenient order. What is the probability that he or she will place the bin of washers in the middle?

There are two ways the assembler can place the bins while keeping the washers in the middle — nuts-washers-bolts, or bolts-washers-nuts. Therefore the numerator of the probability ratio is two.

The denominator is the number of ways three bins can be placed, or the number of permutations of three things taken three at a time.

$$\text{Denominator} = \frac{3!}{(3-3)!}$$

$$= 6$$

The probability of the assembler placing the bin of washers in the middle is

$$p = \frac{2}{6}$$

$$= 1/3, \text{ or } 0.3333$$

Overlapping Probabilities: The Chance of Two or More Events Occurring

This topic finds the probability that two or more different events will all occur. The formula is simply the product of the separate probabilities.

Example 11.15

Your product containers are painted with a random selection of three colors from a bank of twelve different colors. Each container is stamped A, B, C, or D, according to which of four inspectors examine it. (Each inspector examines an equal number.) One container out of six is sent to a customer in Chicago. What is the probability that a container will be painted blue, yellow, and brown, have a B stamped on it, and be sent to Chicago?

First, you look at the probability of the container's being painted blue, yellow, and brown. The denominator of the ratio is the number of combinations of twelve things taken three at a time, or 220. The numerator is one, as there is just one of the 220 combinations that consists of blue, yellow, and brown. Therefore the probability of a container's being painted blue, yellow, and brown is 1/220 or 0.0045.

The probability of having a B stamped on it is 1/4 or 0.25 and the probability of being sent to Chicago is 1/6 or 0.1667.

The probability of one container meeting all three conditions is 0.0045(0.2500)0.1667 or .0002. In general, two containers out of every 10,000 will be blue, yellow, brown, will be stamped B, and will be sent to Chicago.

Mutually Exclusive Probabilities: When There Is No Chance of Both Events Occurring

When the occurrence of one event means another could not possibly occur, the events are mutually exclusive. For example, if a container is painted blue, yellow, and brown, it cannot be painted orange, green, and violet. When the events are mutually exclusive, you can find the probability of one *or* the other by adding their separate probabilities.

Example 11.16

Inspectors reject one item out of 1,500 for exceeding dimensions, and one out of 1,200 for exceeding weight limits. An item rejected for either reason is destroyed immediately, so no item could be rejected for both reasons. What is the probability of an item being rejected?

If it fails either test it is rejected, so the probability of rejection is the sum of the separate probabilities.

$$\text{probability of rejection} = \frac{1}{1,500} + \frac{1}{1,200}$$
$$= 0.0015$$

Binomial Probability

Binomial probability is explained with examples in the quality control chapter (Chapter 6) and a table of binomial probabilities is given as Table A-3 in the appendix.

Briefly, binomial probability applies when there are only two possible results — that is why it is also called *two-state* probability. A production unit can be accepted or rejected; a person can be hired or not; an item can be in stock or out of stock. You cannot use binomial probability if the possibilities include the item being in Stockroom A, Stockroom B, or out of stock.

For a typical demonstration, suppose that, on the average, two percent of the production units are rejected. If you randomly pick four units, binomial calculations can find several probabilities including the probability of one unacceptable unit, the probability of two unacceptable units, the probability of at least three unacceptable units, and so forth.

CHAPTER 12

Forecasting Techniques for Smoothing and Adjusting Data

This chapter looks at some of the forecasting techniques that are applied to the production process, and examines some of the compromises that must be made. Whether it be sales, department budgets, workforce requirements, or something else, your job would be greatly simplified if you knew the future. In the absence of that knowledge, the next best thing to knowing the future is to be able to forecast it with a reasonable degree of accuracy. This chapter shows you how to do that.

There are examples on how to forecast a constant, how to forecast with averages, and how to control your forecasts with a systematic approach known as smoothing. You'll also find out more about trend adjustment and how the trend-smoothing factor works.

FORECASTING AN UNCHANGING QUALITY

Although it may sound like a trivial situation, forecasting an unchanging quality or constant is included for completeness. It is also included as a reminder that, when you use the number, you are making a forecast. As with all forecasts, it is important to understand the conditions under which the forecast is valid.

Example 12.1

The computer that has been in the production department office for two years has been under a maintenance contract that costs $117 every month. The contract guarantees complete maintenance, so there are no deductibles or other costs. How much should you include in next year's budget for computer maintenance?

Twelve times $117 seems reasonable at first. But what if there has been only one service call in those two years? Maybe cancelling the maintenance contract and budgeting for two service calls next year is very conservative.

On the other hand, maybe $117 per month is a bargain that the seller gave us for two years, as an inducement to buy their computer. What will it cost after two years?

Is it certain that the production department is going to keep the computer another year? The point here, and in the rest of the chapter, is to *be wary when using the past to predict the future.*

FORECASTING WITH AVERAGES

Forecasters use past averages as either the forecast itself or as one of the inputs for more complex calculations. To avoid your having to turn back and forth between Chapter 11 and this one, this section covers the topic again—this time arranged specifically to help with forecasting.

There are several types of *average*; each conveys certain information that the others do not; each can be useful when used properly.

Arithmetic Mean

When past data varies

This one is the best-known. In fact, when most people say average, they are thinking of arithmetic mean. It is the simplest method for using the past to predict the future when the past is not a constant. Arithmetic mean is the sum of the data values divided by the number of data values.

$$m = \frac{v_1 + v_2 + v_3 + \cdots + v_n}{n}$$

where

m = arithmetic mean

v_1, v_2, etc = data values number 1, 2, etc.

n = number of data items being evaluated

Example 12.2

The production department monthly telephone expense for five months has been $93, $112, $105, $88, and $101. What amount can you justify in your next budget request?

There doesn't seem to be a trend, either up or down. Unless there is reason to feel that those five months were not typical of any months to come, the arithmetic mean seems like the best figure to use.

$$m = \frac{93 + 112 + 105 + 88 + 101}{5}$$

$$= 99.80$$

You can expect that telephone expenses will be $99.80 for each month in the next budget.

Median

Another type of average is the median, which is the data value that has an equal number of data points above and below it. This type of average is appropriate when the direction in which a data value is off center is more important than how far it is off center.

How to find the median

The procedure for finding the median is to first arrange all the values in numerical order, then locate the midpoint value. If there is an odd number of values, one of them will be the median. If there is an even number, the median will be between two of them; the most widely accepted point is half-way between them (which happens to be their arithmetic mean).

Table 12-1.   Weights of thirty-five orders.

2632	3644	2280	1062	1494	3775	3034
2311	3550	1200	1507	2215	4623	3212
2163	4013	5183	1262	1666	3768	3949
1745	2119	3752	3096	5104	4962	3681
2899	1711	1488	4242	4896	1525	3155

Example 12.3

You are going to design two pallets to handle all your shipments. Table 12-1 lists the weights of shipments for a period that we consider typical. What should be the capacities of the two designs be so that you will use about equal numbers of the pallets?

Here you need the median value—a weight that has just as many above it as below it. The next step is to arrange the list in order of weight, as shown in Table 12-2.

As you have an odd number of data values, the middle one, 3,034 pounds, is the median. One pallet design should be for shipments up to about 3,000 pounds and the other for shipments over 3,000 pounds. Then you can expect to use them in about equal numbers.

Characteristics of the median

It is important to note that the median is not the midpoint between the highest and lowest values. This average is determined simply by the *numbers* of data values.

Unlike arithmetic mean, the median is not affected by actual values, but only by which side of center they are on. If the smallest shipment had been two pounds instead of 1,062, the median would still be 3,034 pounds.

Example 12.4

The forecasters are asked to help with planning work space because the design depends on expected output of individual workers. All ninety-six workers were checked for one hour, with the number of units produced by each worker shown in Table 12-3. Experience shows that this hour is typical, so the data will be used for designing the work space.

Table 12-4 shows the data arranged in order so the forecasters can find the

Table 12-2.   Data arranged in order for finding median.

1062	1200	1262	1488	1494	1507	1525
1666	1711	1745	2119	2163	2215	2280
2311	2632	2899	3034	3096	3155	3212
3550	3644	3681	3752	3768	3775	3949
4013	4242	4623	4896	4962	5104	5183

Table 12-3.   Records of output by ninety-six workers.

60	52	69	37	37	35	48	70	31	65	52	69	45	31	64	68
54	67	67	35	68	41	46	54	58 ·	45	54	70	60	55	47	60
67	58	65	70	35	44	68	52	41	41	60	31	35	31	48	41
40	31	55	33	58	49	35	65	55	53	35	35	33	46	67	31
55	30	35	52	33	30	36	67	33	62	35	61	31	41	54	49
61	33	55	51	31	68	69	65	44	69	32	69	57	40	48	31

median. As the number of data points is even, the median is the arithmetic mean of the forty-eighth and the forty-ninth values.

$$median = (49 + 51)/2$$
$$= 50$$

The work space design should consider that the number of workers producing more than fifty units per hour will equal the number of workers producing less than fifty.

Mode

Another form of average is the mode, which is the value that occurs most often. When the values form a symmetrical, normal distribution, the mean, median, and mode all name the same value. However, when the distribution is skewed, the locations of the three averages can differ significantly.

Example 12.5

Your production department assembles loudspeakers into cabinets, for selling to the audio trade. A manufacturer has offered a very good price on 3,000 loudspeakers of any one size.

If you want to promote a special sale, you will want to select the loudspeaker size that most of your customers want. The last 1,493 sold, which are considered typical, were distributed as shown in Table 12-5.

It does not matter that the mean size sold was 20.89 watts. Neither does it matter that the median was 20 watts. The only measure that is important here is that more customers want 24-watt loudspeakers than any other size. In fact,

Table 12-4.   Production data sorted for finding median.

30	30	31	31	31	31	31	31	31	31	31	32	33	33	33	33
33	35	35	35	35	35	35	35	35	35	36	37	37	40	40	41
41	41	41	41	44	44	45	45	46	46	47	48	48	48	49	49
51	52	52	52	52	53	54	54	54	54	55	55	55	55	55	57
58	58	58	60	60	60	60	61	61	62	64	65	65	65	65	67
67	67	67	67	68	68	68	68	69	69	69	69	69	70	70	70

Table 12-5. Distribution of loudspeaker sales.

4 watt loudspeakers	102 sold
8	98
10	113
12	133
18	127
20	178
24	216
26	150
28	109
30	90
32	58
36	65
40	31
50	23

twenty-one percent more 24-watt units are sold than the median of 20-watt units.

Weighted Average

Some data points might be more significant than others, reducing the value of a straight arithmetic mean. For example, the most recent data might tell more about the future than would data from the distant past. For some forecasts, time lost due to material shortages may be more significant than time lost due to machine breakdowns. Or sales in the Chicago area may reveal more about future sales than do sales in the Los Angeles area.

A weighted average is found by counting certain data values more than others, while dividing by an adjusted number that includes the extra counts. The equation is

$$A_w = \frac{w_1 v_1 + w_2 v_2 + \cdots + w_n v_n}{w_1 + w_2 + \cdots + w_n}$$

where:

A_w = weighted average

w_i = weight assigned to data value number i

v_i = value of data item number i

Example 12.6

Let us assume that, as a forecasting tool, last week's sales are three times as significant as old sales. Sales from two weeks ago are twice as significant. The product has been on the market for ten weeks, with a sales record of 35, 29, 14, 15, 12, 9, 10, 9, 11, and 10. Using a weighted average to forecast next week's sales will give a more realistic forecast because the initial large sales, which no longer occur, have less effect on the calculation.

Weights 1 through 8 are all one, w_9 is two, and w_{10} is three. Substituting these numbers in the formula gives

$$m = \frac{1(35) + 1(29) + 1(14) + 1(15) + 1(12) + 1(9) + 1(10) + 1(9) + 2(11) + 3(10)}{13}$$

$$= 14.23$$

Although the sales seem to have settled at a lower figure, the large sales from ten weeks ago still influence the forecast. Had you taken a straight average, not giving more weight to recent sales, the forecast would have been for 15.40 sales. After a longer period of time, the two weeks of larger sales would have a relatively smaller effect on both the straight average and the weighted average.

Moving Average

The next question is whether sales from a period in the distant past should enter into the forecast at all. A moving average answers that objection by keeping only the most recent data values. Each time a new value is included, the oldest is dropped, so the average is always calculated from a fixed number of values.

The equation for a non-weighted moving average is

$$A_m = \frac{v_{n-i+1} + v_{n-i+2} + \cdots + v_{n-1} + v_n}{i}$$

where:

A_m = moving average over i data values

v_n = value of data unit number n

i = interval, or number of data values to be averaged

This equation tells us to add the last i data values and divide the result by i.

Example 12.7

Labor turnover in the production department has the following history, with turnover in percent:

Year	1	2	3	4	5	6	7	8	9	10
Turnover	3	4	8	7	5	3	4	3	2	3

If you want to find an average to help in forecasting next year's turnover, you should recognize that the reasons for the high rates six and seven years probably do not apply to next year. Therefore, you will use a moving average, and an interval of four years seems reasonable. The forecast for year 11 will be the average of years 7, 8, 9, and 10.

$$f = \frac{4 + 3 + 2 + 3}{4}$$

$$= 3$$

Next year, when you forecast for year 12, you will include the actual value for year 11, and drop the 4 from year 7.

Weighted Moving Average

Even though the older values are being discarded, you may still wish to give more weight to certain ones (such as the very latest) included in the interval. The equation for weighted moving average is:

$$A_{wm} = \frac{w_{n-i+1} \, v_{n-i+1} + \cdots w_{n-1} \, v_{n-1} + w_n \, v_n}{w_{n-i+1} + \cdots w_{n-1} + w_n}$$

where

A_{wm} = weighted moving average

w_n = weight assigned to period n

v_n = value of data unit number n

Example 12.8

Let's use the same sales history used to demonstrate weighted average, and move a four-week interval through the entire period. The latest value in each interval will have a weight of three; the one before it a weight of two, and the other two a weight of one.

The first weighted moving average, for years one through four, is

$$A_{wm1} = \frac{1(35) + 1(29) + 2(14) + 3(15)}{1 + 1 + 2 + 3}$$

$$= 19.57$$

Just for comparison, the straight average of these four weeks is 23.25. After five weeks, you make the following changes:

discard the thirty-five from week 1

change the weight of week 3 to 1

change the weight of week 4 to 2

Table 12-6. Weighted moving average.

Interval	Weighted moving average
1–4	19.57
2–5	15.57
3–6	11.43
4–7	10.71
5–8	9.71
6–9	10.00
7–10	10.14

include twelve (week 5) with a weight of three

$$A_{wm2} = \frac{1(29) + 1(14) + 2(15) + 3(12)}{1 + 1 + 2 + 3}$$

$$= 15.57$$

Table 12-6 shows a four-week interval moving through these ten weeks, with the latest always having a weight of three; the second from the latest always has a weight of two.

In choosing how many items to include in the moving interval, note that the larger intervals build a bridge that diminishes the effects of short transient changes. On the other hand, the longer the interval, the longer it will take for a new trend to be recognized. The next section shows systematic ways to smooth the results.

SMOOTHING THE RESULTS: A SYSTEMATIC WAY TO COMPROMISE YOUR FORECASTS

Smoothing *always* requires a compromise — a high degree of smoothing reduces the forecast's reaction to transients, but delays recognition of permanent changes. It has some of the same characteristics as the long run average.

Very little smoothing is about the same as using the previous period's results as a forecast. The forecast will follow transients closer.

Exponential Smoothing

Books on the subject have favorite versions of the formula, but they all have one thing in common. They all add a correction that is based on the difference between forecasts and actual results of one or more past periods. Our formula algebraically adds a percentage of the difference between the preceding period's forecast and its actual result.

$$f_n = f_{n-1} + s(a_{n-1} - f_{n-1})$$

where

f_n = smoothed forecast for period n

f_{n-1} = forecast for preceding period

a_{n-1} = actual results of preceding period

s = smoothing factor between 0 and 1

The last term finds whether the previous forecast was high or low, then adds (algebraically) a portion of that error.

Examining the Variables

Because of the many interpretations given to this formula, some further explanation will be helpful. First, note that, as used here, a *small* value of smoothing factor gives a *high* degree of smoothing. A larger value of s causes the forecast to closely follow actual results.

Some books consider that, when making a new forecast, the last forecast calculated is locked in to become f_{n-1}. We consider the forecast to be more realistic if f_{n-1} is an adjusted figure, as will be shown in an example shortly. The adjustment is not the same as the correction that is made because of the error between forecast and actual figures.

Example 12.9

Table 12-7 shows twenty periods of sales that exhibit a steady rise through the tenth period, then a steady decline through the nineteenth. One exception to this pattern is the one-period shock in the fifteenth period — sales of 1,000 when the pattern indicates 500. The purpose of that transient is to examine the effects of different amounts of smoothing.

Calculations with three different values of smoothing, with a moving interval of three, are shown in Table 12-8. Before examining the results, it is important to understand the exact meaning of the numbers in each column.

Table 12-7.   Sales record for example of smoothing.

Period	Sales	Period	Sales
1	100	11	900
2	200	12	800
3	300	13	700
4	400	14	600
5	500	15	1000
6	600	16	400
7	700	17	300
8	800	18	200
9	900	19	100
10	1000	20	200

Table 12-8. Calculations for comparison of smoothing factor.

Period	Actual	Average to date	Un-smoothed	Moving average over 3 periods Smoothed with .1	Smoothed with .3	Smoothed with .5
1	100	100	—	—	—	—
2	200	150	—	—	—	—
3	300	200	200	—	—	—
4	400	250	300	210	230	250
5	500	300	400	310	330	350
6	600	350	500	410	430	450
7	700	400	600	510	530	550
8	800	450	700	610	630	650
9	900	500	800	710	730	750
10	1000	550	900	810	830	850
11	900	582	933	910	930	950
12	800	600	900	930	923	917
13	700	608	800	890	870	850
14	600	607	700	790	770	750
15	1000	633	767	690	670	650
16	400	619	667	790	837	883
17	300	600	567	640	587	533
18	200	578	300	540	487	433
19	100	553	200	290	270	250
20	200	535	167	190	170	150
21				170	177	183

Average to date

Average to date is the long-run average. It is simply the sum of actual sales in all preceding periods, plus the present period, divided by the number of the present period. Obviously, this information is not known until the present period is completed so, if it is to be used as a forecast, it should forecast the following period. That is, the figure shown for period five (300) would be the forecast for period six.

Unsmoothed moving average

Unsmoothed moving average is also an after-the-fact figure. As it averages the latest three months, it is not calculated for the first two months of operation. After two months, the figure shown by any period can be used as the forecast for the next period. That is, the three-month moving average for period eight (the average of periods six, seven, and eight — 700) can be used as the forecast for period nine.

Smoothed moving average

Smoothed moving average columns use the unsmoothed moving average through the preceding period as f_{n-1}. For example, the smoothed forecasts for period nine were calculated as follows:

$$f_9 = \begin{array}{c} \text{average of} \\ \text{actuals for} \\ \text{periods 6, 7, 8} \end{array} + s \left[\begin{array}{ccc} \text{actual} & & \text{average of} \\ \text{sales for} & - & \text{actuals for} \\ \text{period 8} & & \text{periods 6, 7, 8} \end{array} \right]$$

The calculation with a smoothing factor of 0.1 used the following numbers:

$$f_9 = \frac{600 + 700 + 800}{3} + 0.1 \left[800 - \frac{600 + 700 + 800}{3} \right]$$
$$= 710$$

Figure 12-1 makes the results clearer. First, note the graph of average to date, which is reluctant to reflect new data. During the rise in sales through period ten, this forecast becomes ever further from the actual values because the low initial values are always present in the calculation. This forecast barely acknowledges the transient in period fifteen.

The forecasts based on three-month moving averages show that, as s increases, the forecast is faster to reflect actual changes. When sales began their long decline, the forecast using $s = 0.5$ quickly became the closet to actual values.

On the other hand, $s = 0.5$ caused the forecast to respond so quickly to the transient that it told us to prepare to sell 883 in period sixteen. A smoothing factor of 0.1 is more of a wait-and-see factor; it forecast sales of 790 after the transient.

In the twentieth period the forecast smoothed with 0.1 is very close to the actual value. However, that happened because the actual value suddenly turned around and ran into the forecast. The situation is similar to a clock that loses one second a day; after being set and left alone, it won't show the correct time again for almost 120 years — but a clock that's stopped shows the correct time twice a day.

The conclusion is that no value of smoothing is categorically best. One workable policy would be to use a small s during regular times, when period-to-period changes are likely to be transient. After a major change in factors that influence sales, such as introduction of a "new improved" version of the product or an advertising campaign stressing brand loyalty, when increased sales are expected to continue, use a large smoothing factor.

COMPUTER PROGRAM FOR FORECAST SMOOTHING

Several variables are involved, and repeated calculations are necessary each time a variable value is changed. The computer program listed in Figure 12-2 makes any number of forecasts in a matter of seconds. It can be used to experiment

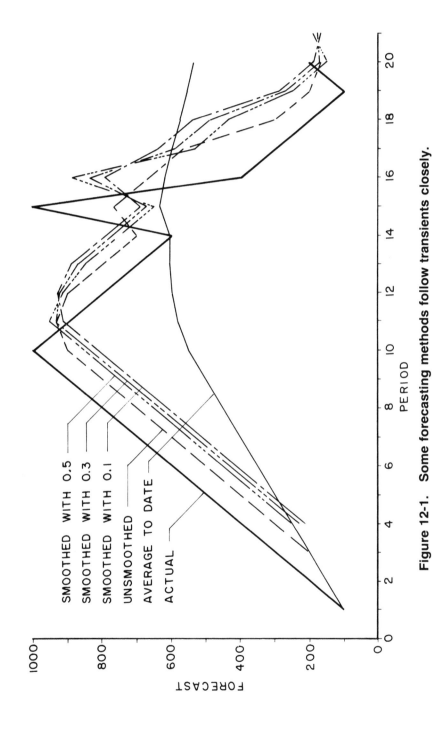

Figure 12-1.   Some forecasting methods follow transients closely.

with different smoothing factors and moving interval lengths, or to make professional forecasts.

Although that program was tested on an IBM PC, most of the lines are standard BASIC that will be understood by other computers.

Example 12.10

Investigate the effect of different moving average periods on a sales function that holds steady for a few periods, then rises in a straight line, then holds steady at a new level. The set of data values given below meets these conditions.

RUN the program and it will stop with the menu on the screen. Take choice number one, which allows you to enter any number of new data values from the keyboard. The screen will then ask you for data value #1. Type **100** and press **Enter**. It will ask for data value #2. Continue responding with the following values and press **Enter** after each response:

data value #2 100	data value #10 500
3 100	11 500
4 100	12 500
5 200	13 500
6 300	14 500
7 400	15 500
8 500	16 500
9 500	

When the screen asks for #17, type **END**. The menu will reappear; select #4 for the calculations.

Now the program gives you the chance to experiment. Table 12-9 shows the results with smoothing factors of .1, .3, and .5, and moving average intervals of 3. As often as you want, you can return to the menu, select #4, and have the computer make the calculations with another set of numbers. When the screen asks for three smoothing factors, be sure to separate them with commas, as shown in Table 12-9. Do not press **Enter** until you type all three of them.

When you want to enter a new set of data values, select #1 from the menu.

SMOOTHING WITH TREND ADJUSTMENT

Figure 12-1 shows that all the forecasts lag when the actual values settle down to a pattern of steady increases or decreases. You can improve the forecast by adding a trend correction. That is, to the exponentially smoothed forecast, add a fraction of the latest change plus a fraction of the average of the latest changes. The literature contains many variations on the concept of adjusting to trends; the one presented here is among the simplest. Some formulas get into smoothing with trends of trends. This simple formula is a compromise between simplicity (which sacrifices detail) and complexity (which gives more detail, but at the expense of making calculations more difficult).

Figure 12-2. Computer program calculating forecasts with smoothing.

```
10 '*******************************************************************
20 '** SMTHDEM.BAS  Demonstration of effect of smoothing factor  **
30 '** for forecasting chapter of Handbook of Manufacturing and  **
40 '** Production Formulas, Charts, and Tables                    **
50 '*******************************************************************
60 CLEAR,,,32768!
70 SCREEN 6: COLOR 3,2
80 U$="####"
90 DIM V(100),OAAVG(100),F1(100),F3(100),F5(100),MAVG(100)
100 KEY OFF
110 CLS
120 PRINT " 1 Enter new set of data from keyboard"
130 PRINT " 2 Retrieve set of data from disk"
140 PRINT " 3 Save set of data on disk"
150 PRINT
160 PRINT " 4 Calculate and display general table"
170 PRINT
180 INPUT "Select by number ",SELECT
190 IF SELECT>0 AND SELECT<5 THEN 210
200 PRINT "Select between 1 and 4 please.": GOTO 120
210 CLS
220 ON SELECT GOSUB 260,340,420,500
230 GOTO 110
240 END
250 'Subroutine to enter new data from keyboard  111111111111111
260 N=0
270 PRINT "Enter data value #";N+1;"  .... or END";
280 INPUT " ",A$
290 IF A$="END" OR A$="end" THEN RETURN
300 N=N+1
310 V(N)=VAL(A$)
320 GOTO 270
330 'Subroutine to retrieve data from disk   2222222222222222
340 N=0
350 INPUT "Name of file to be retrieved ",N$
360 OPEN N$ FOR INPUT AS #1
370 IF EOF(1) THEN CLOSE: RETURN
380 N=N+1
390 INPUT #1, V(N)
400 GOTO 370
410 'Subroutine to save data in memory  3333333333333333
420 INPUT "Name to be used for file ",N$
430 OPEN N$ FOR OUTPUT AS #3
440 FOR C=1 TO N
450    WRITE #3, V(C)
460 NEXT
470 CLOSE
480 RETURN
490 'Subroutine to calculate general table  44444444444444
500 SUMALL=0: MSUM=0
510 GOSUB 680
520 FOR C=1 TO N+1
530    IF C>N THEN 610
540    SUMALL=SUMALL+V(C)
550    OAAVG(C)=SUMALL/C
560    MSUM=MSUM+V(C)
570    IF C<M THEN 600
580    MSUM=MSUM-V(C-M)
590    MAVG(C)=MSUM/M
600    IF C<M+1 THEN 640
```

Figure 12-2. (continued)

```
610    F1(C)=MAVG(C-1)+S1*(V(C-1)-MAVG(C-1))
620    F3(C)=MAVG(C-1)+S2*(V(C-1)-MAVG(C-1))
630    F5(C)=MAVG(C-1)+S3*(V(C-1)-MAVG(C-1))
640 NEXT
650 GOSUB 750
660 RETURN
670 'Subroutine for operator input of certain factors  <<<<<<<<<<<<<<
680 INPUT "Enter three values of smoothing factor ",S1,S2,S3
690 IF S1>=0 AND S2>=0 AND S3>=0 AND S1<=1 AND S2<=1 AND S3<=1 THEN 720
700 PRINT "Smoothing factor must between 0 and 1"
710 GOTO 680
720 INPUT "How many periods should be included in moving average? ",M
730 RETURN
740 'Subroutine to print table  <<<<<<<<<<<<<<<<<<
750 PRINT TAB(35)"Moving average over";M;"periods"
760 PRINT TAB(19)"Average";TAB(31)"Un-";TAB(40)"Smoothed";TAB(51)"Smoothed";TAB(
62)"Smoothed"
770 PRINT "Period";TAB(10)"Actual";TAB(19)"to date";TAB(29)"smoothed";TAB(40)"wi
th";S1;TAB(51)"with";S2;TAB(62)"with";S3
780 FOR C=1 TO N+1
790    PRINT USING U$;C;
800    IF C>N THEN 900
810    PRINT TAB(12)USING U$;V(C);
820    PRINT TAB(21) USING U$; OAAVG(C);
830    IF C>M THEN 860
840    IF C=M THEN PRINT TAB(31) USING U$; MAVG(C);:GOTO 860
850    PRINT TAB(32)"-";
860    IF C>=M+1 THEN 890
870    PRINT TAB(43)"-";TAB(54)"-";TAB(65)"-"
880    GOTO 930
890    PRINT TAB(31) USING U$; MAVG(C);
900    PRINT TAB(42) USING U$; F1(C);
910    PRINT TAB(53) USING U$; F3(C);
920    PRINT TAB(64) USING U$; F5(C)
930 NEXT
940 LOCATE 25,1
950 INPUT "Press Enter to return to menu ",E$
960 RETURN
```

$$\begin{array}{l}\text{forecast}\\\text{with trend} \\\text{adjustment}\end{array} = \begin{array}{l}\text{exponentially}\\\text{smoothed}\\\text{forecast}\end{array} + \left[\begin{array}{l}\text{trend}\\\text{smoothing}\\\text{factor}\end{array}\right]\begin{array}{l}\text{latest}\\\text{trend} +\end{array} \left[\begin{array}{l}1-\text{trend}\\\text{smoothing}\\\text{factor}\end{array}\right]\begin{array}{l}\text{trend}\\\text{moving}\\\text{average}\end{array}$$

$$f_t = f_e + s_t\,(t_l) + (1 - s_t)(t_a)$$

where

f_e is the exponentially smoothed factor demonstrated in the preceding section

s_t is a smoothing factor that determines the degree to which the latest trend affects the forecast. This factor may or may not have the same value as the smoothing factor used for exponential smoothing.

t_l is the trend from two periods back to one period back — the difference between actual results for those periods.

Table 12-9. Example of use of computer program to investigate effect of smoothing factor value.

Enter three values of smoothing factor .1,.3,.5
How many periods should be included in moving average? 3

Period	Actual	Average to date	Un-smoothed	Moving average over 3 periods Smoothed with .1	Smoothed with .3	Smoothed with .5
1	100	100	—	—	—	—
2	100	100	—	—	—	—
3	100	100	100	—	—	—
4	100	100	100	100	100	100
5	200	120	133	100	100	100
6	300	150	200	140	153	167
7	400	186	300	210	230	250
8	500	225	400	310	330	350
9	500	256	467	410	430	450
10	500	280	500	470	477	483
11	500	300	500	500	500	500
12	500	317	500	500	500	500
13	500	331	500	500	500	500
14	500	343	500	500	500	500
15	500	353	500	500	500	500
16	500	363	500	500	500	500
17				500	500	500

t_a is the latest moving average of the trends, using an interval that may or may not be the same as the interval used in calculating the exponentially smoothed forecast.

The coefficients of the last two terms (s_t and $1 - s_t$) control the distribution of influence between the very latest change (trend) and the average of the last few changes.

Example 12.11

Using the sales data from the example with the transient in period fifteen, make forecasts using a moving average interval of 3 for both the exponentially smoothed term and the trends. Use smoothing factors of 0.3 for the exponential and the trend smoothing respectively.

To determine which period can be the first one forecast, note that you want a moving average of trends using a sliding interval of three. That means that the first average that can be calculated will be the average of the trends from period one to period two, from period two to period three, and from period three to period four. Therefore, no forecast can be made until the actual results from periods one through four are known; the first forecast will be for period five.

Table 12-8 forecasts 330 for period five with a smoothing factor of 0.3. That forecast now becomes f_e in the new equation.

In the second term, the latest trend equals the actual result for period four minus the actual result of period three. Multiplying that difference by the smoothing factor for trends (0.5) gives

$$\text{second term} = 0.5(400 - 300)$$

$$= 50$$

For the last term you need the average of the first three trends. From period one to period two the trend is $200 - 100$, or 100; from period two to period three it is $300 - 200$, or 100; from period three to period four it is $400 - 300$, or 100. The average of these three trends is $(100 + 100 + 100)/3$, or 100. Therefore, the last term is

$$\text{last term} = (1 - 0.5)(100)$$

$$= 50$$

Combining the three terms gives the forecast for period five.

$$f_5 = 330 + 50 + 50$$

$$= 430$$

Table 12-10 shows some of the partial calculations and the rest of the forecasts through period twenty-one.

Table 12-10. Forecasts and partial calculations with trend adjustment.

Period	Actual	Trend of actual values Back 1	Back 2	Back 3	Mov avg trend	Smoothed trend	Forecast
1	100						
2	200						
3	300						
4	400						
5	500	100	100	100	100	100	430
6	600	100	100	100	100	100	530
7	700	100	100	100	100	100	630
8	800	100	100	100	100	100	730
9	900	100	100	100	100	100	830
10	1000	100	100	100	100	100	930
11	900	100	100	100	100	100	1030
12	800	−100	100	100	33	−33	890
13	700	−100	−100	100	−33	−67	803
14	600	−100	−100	−100	−100	−100	670
15	1000	−100	−100	−100	−100	−100	570
16	400	400	−100	−100	67	233	1070
17	300	−600	400	−100	−100	−350	237
18	200	−100	−600	400	−100	−100	387
19	100	−100	−100	−600	−267	−183	87
20	200	−100	−100	−100	−100	−100	70
21		100	−100	−100	−33	33	210

Figure 12-3.   Computer program forecasts with trend adjustment.

```
10 '*****************************************************************
20 '** FCSTTR.BAS  Forecast calculation with smoothed trend    **
30 '** for Handbook of Manufacturing and Production Management **
40 '** Formulas, Charts, and Tables                            **
50 '*****************************************************************
60 CLEAR,,,32768!
70 SCREEN 6: COLOR 3,1
80 U$="####"
90 DIM V(100),F1(100),F2(100),MAVG(100),T(100),TMOV(100),SMTREND(100)
100 KEY OFF
110 CLS
120 PRINT " 1 Enter new set of data from keyboard"
130 PRINT " 2 Retrieve set of data from disk"
140 PRINT " 3 Save set of data on disk"
150 PRINT
160 PRINT " 4 Calculate and display general table"
170 PRINT
180 INPUT "Select by number ",SELECT
190 IF SELECT>0 AND SELECT<5 THEN 210
200 PRINT "Select between 1 and 4 please.": GOTO 120
210 CLS
220 ON SELECT GOSUB 260,340,420,500
230 GOTO 110
240 END
250 'Subroutine to enter new data from keyboard  111111111111111
260 N=0
270 PRINT "Enter data value #";N+1;"   .... or END";
280 INPUT " ",A$
290 IF A$="END" OR A$="end" THEN RETURN
300 N=N+1
310 V(N)=VAL(A$)
320 GOTO 270
330 'Subroutine to retrieve data from disk   2222222222222222
340 N=0
350 INPUT "Name of file to be retrieved ",N$
360 OPEN N$ FOR INPUT AS #1
370 IF EOF(1) THEN CLOSE: RETURN
380 N=N+1
390 INPUT #1, V(N)
400 GOTO 370
410 'Subroutine to save data in memory  3333333333333333
420 INPUT "Name to be used for file ",N$
430 OPEN N$ FOR OUTPUT AS #3
440 FOR C=1 TO N
450    WRITE #3, V(C)
460 NEXT
470 CLOSE
480 RETURN
490 'Subroutine to coordinate calculations and table  444444444444444
500 GOSUB 570
510 GOSUB 810
520 GOSUB 930
530 LOCATE 25,1
540 INPUT "Press Enter for menu ",E$
550 GOTO 110
560 'Subroutine to calculate smoothed trend  <<<<<<<<<<<<<<<<
570 INPUT "How many periods should be included in moving average? ",M
580 IF M>0 AND M<N+1 THEN 620
590 PRINT "Number of periods can't be larger than number"
600 PRINT "of data values!!!"
```

Figure 12-3. (continued)

```
610 GOTO 570
620 INPUT "Enter smoothing factor for exponential smoothing ",SE
630 IF SE>=0 AND SE<=1 THEN 660
640 PRINT "Smoothing factor must be between 0 and 1"
650 GOTO 620
660 INPUT "Enter smoothing factor for trend ",ST
670 IF SL>=0 AND SL<=1 THEN 700
680 PRINT "Smoothing factor must be between 0 and 1"
690 GOTO 660
700 FOR C=M+2 TO N+1
710 SUMT=0
720   FOR MOVPER=C-1 TO C-M STEP -1
730     T(MOVPER)=V(MOVPER)-V(MOVPER-1)
740     SUMT=SUMT+T(MOVPER)
750   NEXT MOVPER
760   TMOV(C-1)=SUMT/M
770   SMTREND(C)=ST*(V(C-1)-V(C-2))+(1-ST)*TMOV(C-1)
780 NEXT C
790 RETURN
800 'Subroutine to calculate forecast with exponential smoothing  <<<<<<<<<
810 MSUM=0
820 FOR C=1 TO N+1
830   IF C>N THEN 890
840   MSUM=MSUM+V(C)
850   IF C<M THEN 880
860   MSUM=MSUM-V(C-M)
870   MAVG(C)=MSUM/M
880   IF C<M+1 THEN 900
890   F1(C)=MAVG(C-1)+SE*(V(C-1)-MAVG(C-1))
900 NEXT C
910 RETURN
920 'Subroutine to finish forecast and print table  <<<<<<<<<<<<<
930 FOR C=1 TO N+1
940   F2(C)=F1(C)+SMTREND(C)
950 NEXT C
960 PRINT TAB(19)"Trend of actual values";TAB(44)"Mov avg";TAB(54)"Smoothed"
970 PRINT "Period";TAB(10)"Actual";TAB(19)"Back 1";TAB(27)"Back 2";TAB(35)"Back
3";TAB(45)"trend";TAB(55)"trend";TAB(65)"Forecast"
980 FOR C=1 TO N+1
990   PRINT USING U$;C;
1000   IF C>N THEN 1020
1010   PRINT TAB(10) USING U$;V(C);
1020   IF C<M+2 THEN PRINT: GOTO 1090
1030   PRINT TAB(21) USING U$;V(C-1)-V(C-2);
1040   PRINT TAB(29) USING U$;V(C-2)-V(C-3);
1050   PRINT TAB(37) USING U$;V(C-3)-V(C-4);
1060   PRINT TAB(47) USING U$;TMOV(C-1);
1070   PRINT TAB(57) USING U$;SMTREND(C);
1080   PRINT TAB(67) USING U$;F2(C)
1090 NEXT C
1100 RETURN
```

Trend Smoothing Factor

As books of the subject offer various formulas that smooth the effect of trend in different ways, the factor should be examined. Some formulas use the smoothing factor to control the total influence of trends; here it is used to control the relative influence between the lastest trend and the moving average trend. The total influence of trends is always a factor of one because the coeffi-

Table 12-11. Results with trend smoothing factor of 0.2.

Period	Actual	Trend of actual values Back 1	Back 2	Back 3	Mov avg trend	Smoothed trend	Forecast
1	100						
2	200						
3	300						
4	400						
5	500	100	100	100	100	100	430
6	600	100	100	100	100	100	530
7	700	100	100	100	100	100	630
8	800	100	100	100	100	100	730
9	900	100	100	100	100	100	830
10	1000	100	100	100	100	100	930
11	900	100	100	100	100	100	1030
12	800	−100	100	100	33	7	930
13	700	−100	−100	100	−33	−47	823
14	600	−100	−100	−100	−100	−100	670
15	1000	−100	−100	−100	−100	−100	570
16	400	400	−100	−100	67	133	970
17	300	−600	400	−100	−100	−200	387
18	200	−100	−600	400	−100	−100	387
19	100	−100	−100	−600	−267	−233	37
20	200	−100	−100	−100	−100	−100	70
21		100	−100	−100	−33	−7	170

cients are s and $1 - s$. If s is large (near one), the latest trend is more effective and $1 - s$ approaches zero, making the average trend less effective.

COMPUTER PROGRAM FOR TREND ADJUSTMENT FORECASTS

One of the features of these formulas is the flexibility they allow for choosing how forecasts are to react to changes and trends. However, the calculations become tiresome when repeated for comparisons and when making a long table. The computer program listed in Figure 12-3 makes investigations of any size easy. Operator responses, such as moving average interval, are clearly requested on the screen.

Once a set of data is in the computer's memory (either new data from the keyboard or old data from a disk file), unlimited calculations can be made with various combinations of variable values. Each time the menu appears and you choose calculations, you have an opportunity to change the variables.

Example 12.12

Compare the effect of emphasizing the latest trend versus emphasizing a moving average of trends.

Let's make two sets of calculations for the comparison — one with a trend

Table 12-12.   Results with trend smoothing factor of 0.8.

Period	Actual	Trend of actual values Back 1	Back 2	Back 3	Mov avg trend	Smoothed trend	Forecast
1	100						
2	200						
3	300						
4	400						
5	500	100	100	100	100	100	430
6	600	100	100	100	100	100	530
7	700	100	100	100	100	100	630
8	800	100	100	100	100	100	730
9	900	100	100	100	100	100	830
10	1000	100	100	100	100	100	930
11	900	100	100	100	100	100	1030
12	800	−100	100	100	33	−73	850
13	700	−100	−100	100	−33	−87	783
14	600	−100	−100	−100	−100	−100	670
15	1000	−100	−100	−100	−100	−100	570
16	400	400	−100	−100	67	333	1170
17	300	−600	400	−100	−100	−500	87
18	200	−100	−600	400	−100	−100	387
19	100	−100	−100	−600	−267	−133	137
20	200	−100	−100	−100	−100	−100	70
21		100	−100	−100	−33	73	250

smoothing factor of 0.2 and one with 0.8. Other variables will be held constant so as not to confuse the comparison.

RUN the program and then enter the set of data that has a transient in period fifteen. Choose a moving average interval of three and an exponential smoothing factor of 0.3. Then respond to "Enter smoothing factor for trend" with .2. When the table appears on the screen, copy it or transfer it to your printer if you have one connected. Then return to the menu, choose calculations again, and enter the same first two responses. Enter .8 for the trend smoothing factor.

The results are shown in Tables 12-11 and 12-12. As the latest trend and the average trend are the same through period ten, the forecasts through period eleven are the same. But then notice how violently the forecast reacts (overreacts) to the transient when s_t is 0.8, emphasizing the latest trend. The same overreaction occurs after actual sales return to normal in period sixteen.

MATHEMATICAL REGRESSION

With mathematical regression, you can determine the equation that best fits a given set of data. The topic is covered thoroughly in Chapter 13, with some examples that demonstrate forecasting.

For the present chapter, consider that mathematical regression is another technique for obtaining the basic forecast, instead of using a moving average. Then smoothing and trend adjustment can be applied.

CONCLUSION

With one exception these forecasts have been based on nothing except the history of actual results. The one exception was to change the smoothing factor when a new level of sales is expected to continue, as when a new improved version is introduced. It is expected that other conditions, when known, also be used to modify the forecast. However, the calculated results should be the main part of the forecast, with judgment factors applied from there.

No formula based only on the history of results can forecast a change such as occurs in period fifteen. It can react only when period fifteen becomes part of the history. One way to improve the forecasts is to record results for shorter periods — quarters or months if possible. Then it is more likely that the transient would be detected in a build-up.

CHAPTER 13

Mathematical Regression: How to Find the Best Equation to Fit Your Data

Although you may often make a best-fit estimate by laying a ruler through the data points on a graph, you will seldom find the best fit that way. It is easy for a pattern of data points to form an optical illusion that makes us place the best fit line wrong. This chapter gives you the equation that is always mathematically the best fit.

After learning what "best fit" means in mathematical regression, you'll find examples of the straight line process known as linear regression, to help you forecast, simulate, and otherwise test the effects of a particular policy you may be thinking of implementing. Also discussed is moving regression and how to get the best fit equation for data in an interval that moves through the data. There is also a computer program for calculating the regression equations.

HOW YOU CAN USE MATHEMATICAL REGRESSION

The objective of mathematical regression is to find an equation that describes an operation, so that you can use the equation to forecast, simulate, or otherwise test the effects of policies you are considering. For example, suppose your project makes balloons for weather research, and you have inflated several balloons to different pressures. For each pressure, you have measured the altitude the balloon reaches. With the techniques of this chapter, you can turn the data into an equation that will allow you to predict the most probable altitude a balloon will reach for any given pressure.

BEST FIT: WHAT IT MEANS IN MATHEMATICAL REGRESSION

Early in algebra we learned to draw the graph of an equation. Regression takes us in the other direction; we find an equation that best fits a set of data points.

Start with a set of data points on a graph, and draw a line through them. Then measure the vertical distance from each point to the line. As ideally the line would pass right through each point, this vertical distance is the error of each point. Some errors will be positive and some negative because some points are above the line and some below.

One more step is necessary, because the positive and negative errors cancel, either partially or completely. The last step is to square each error, so they will all be positive, and then add all the squared values.

You could repeat the process with other lines through the scatter of data points on the graph. For each line you could calculate the sum of squared errors. The line with the lowest sum of squared errors is the best fit, and is often called the *least squares best fit* line.

You don't actually draw lines, calculate errors, and try to find the best line experimentally. This chapter gives you a mathematical procedure that takes you directly to the equation of the best line.

LINEAR REGRESSION: A STRAIGHT-LINE PROCESS

There are several types of lines that can be fit to the data. This chapter is restricted to linear regression, which means that you fit the best *straight* line. Because of its simplicity, linear regression is often used even when it is clear that the data points form a curve. Inaccuracies due to fitting a straight line to curved data can be reduced by treating the data as a series of segments that approximate straight lines. An example of that procedure is given later in the chapter.

EQUATION OF STRAIGHT LINE

A straight line can always be an equation in the form

$$y = mx + b$$

where m is the slope (difference in ys divided by difference in xs over any segment of the line) and b is the value of y when x equals zero (the point where the line crosses the Y axis). Thus

$$y = 5x + 3$$

$$y = 2x - 7$$

$$y = -4x + 8$$

$$y = \tfrac{6}{7}x + 12$$

are all equations of straight lines. When the slope has a negative value, as in the third equation, the line slants so that its left side is higher than its right side; a line with positive slope slants with its right side higher. The objective of linear regression is to calculate values for m and b.

Regression Equations

The equations for calculating m and b are

$$m = \frac{n\Sigma xy - \Sigma x \Sigma y}{n\Sigma x^2 - (\Sigma x)^2}$$

$$b = \frac{\Sigma y - m\Sigma x}{n}$$

where

n = the number of data points

Σxy = the sum of xy products of each data point

Σx = the sum of all the x values

Σy = the sum of all the y values

Σx^2 = the sum of the squares of all the x values

Table 13-1.   Machine breakdown records for six years.

Year	Average weekly breakdowns
1	4.2
2	4.8
3	5.1
4	5.7
5	6.3
6	6.9

Notice that m is used in calculating b and therefore must be calculated first.

Times Series

Example 13.1

Let's forecast average weekly machine breakdowns, based on records from the past six years, given in Table 13-1. Consider the years to be x values and the breakdowns to be ys. It is best to organize the work in a table, as shown in Table 13-2.

Then, substituting sums and n ($n = 6$) in the equations gives

$$m = \frac{6(124.8) - 21(33)}{6(91) - 91}$$

$$= 0.531$$

$$b = \frac{33 - 0.531(21)}{6}$$

$$= 3.64$$

The line that best fits the data is therefore found from the equation

$$y = 0.531(\text{year number}) + 3.64$$

If the same machines are used for another year (the seventh year), the following prediction can be made

$$\text{Average weekly breakdowns} = 0.531(7) + 3.64$$

$$= 7.36$$

Table 13-2.   Calculations organized in a table.

x	y	x^2	xy
1.0	4.2	1.0	4.2
2.0	4.8	4.0	9.6
3.0	5.1	9.0	15.3
4.0	5.7	16.0	22.8
5.0	6.3	25.0	31.5
6.0	6.9	36.0	41.4
21.0	33.0	91.0	124.8

Series not based on time

Example 13.1 showed a *time series*, so named because the x axis is a measure of time, and the data items are listed chronologically. There is only one y value paired with each value of time. In this example there was only value of average weekly breakdowns for each year.

Regression works equally well with relationships other than time. For example, records could be kept on percent of output rejected by inspectors as a function of number produced daily. Then a person plotting the points on a graph would not move smoothly from left to right; he or she might plot six rejects for a 1,000 unit day, then four rejects for a 790 unit day, then back up to 1,050 unit day with five rejects, and so forth. The formulas for *m* and *b* tell us that the order in which data is used in the calculations is not important; all that is needed is the sums (of the *x* values, and the like).

Example 13.2

Applicants for employment on the assembly line are given a test whose scores can range from ten to seventy. After they are thoroughly familiar with the work, their production ratings are recorded. We will use the results to develop a formula that will help us predict, after the test, how productive the employee will be.

Table 13-3 shows the records of fifty employees, not arranged in any special order. Test scores will be *x* values and productivity ratings will be the *y*s.

Table 13-4 shows the calculations, leading to the equation

$$\text{production rating} = 2.122(\text{test score}) + 5.425$$

That equation can now be used for predicting the productivity rating of applicants; the test can help place employees on jobs for which they are best suited.

For example, if an applicant scores 40 on the test, the equation shows

$$\text{production rating} = 2.122(40 + 5.425$$

$$= \text{about } 90$$

Not every applicant who scores 40 will settle down to a productivity rating of 90. The purpose of finding the regression equation is to calculate values that are likely but not guaranteed. Note also that any particular regression equation is based on data from a specific group; a prediction made from the equation assumes that past data is typical of future data.

As can be seen from Table 13-3, calculating the regression equation can be time-consuming, even when using a calculator. To simplify the task, a computer program is provided later in the chapter.

MOVING REGRESSION: HOW TO GET THE BEST-FIT EQUATION FOR DATA IN AN INTERVAL

The chapter on forecasting looked at moving average — the average of values in an interval that moves through the data. Moving regression works in the same

Table 13-3. Test scores paired with productivity ratings.

	x	y
1	22.00	57.24
2	20.00	40.00
3	32.00	66.60
4	69.00	133.20
5	16.00	33.60
6	44.00	88.20
7	31.00	57.60
8	14.00	41.80
9	63.00	149.60
10	30.00	77.00
11	65.00	168.00
12	30.00	71.40
13	46.00	91.80
14	56.00	134.20
15	41.00	91.08
16	37.00	75.60
17	19.00	38.40
18	66.00	127.80
19	34.00	93.60
20	61.00	145.20
21	17.00	35.20
22	45.00	92.00
23	43.00	86.40
24	50.00	101.20
25	63.00	176.80
26	27.00	70.40
27	52.00	123.12
28	60.00	117.00
29	41.00	82.80
30	30.00	73.50
31	28.00	59.40
32	11.00	25.60
33	64.00	133.86
34	57.00	148.80
35	46.00	112.20
36	50.00	101.20
37	48.00	116.60
38	33.00	83.60
39	46.00	112.20
40	64.00	143.52
41	13.00	28.80
42	54.00	108.56
43	70.00	120.00
44	49.00	118.80
45	28.00	72.60
46	14.00	41.80
47	27.00	51.20
48	46.00	122.40
49	22.00	64.80
50	39.00	79.20

Table 13-4. Calculations to find regression equation.

x	y	x sq	xy
22	57.24	484	1259.28
20	40.00	400	800.00
32	66.60	1024	2131.20
69	133.20	4761	9190.80
16	33.60	256	537.60
44	88.20	1936	3880.80
31	57.60	961	1785.60
14	41.80	196	585.20
63	149.60	3969	9424.80
30	77.00	900	2310.00
65	168.00	4225	10920.00
30	71.40	900	2142.00
46	91.80	2116	4222.80
56	134.20	3136	7515.20
41	91.08	1681	3734.28
37	75.60	1369	2797.20
19	38.40	361	729.60
66	127.80	4356	8434.80
34	93.60	1156	3182.40
61	145.20	3721	8857.20
17	35.20	289	598.40
45	92.00	2025	4140.00
43	86.40	1849	3715.20
50	101.20	2500	5060.00
63	176.80	3969	11138.40
27	70.40	729	1900.80
52	123.12	2704	6402.24
60	117.00	3600	7020.00
41	82.80	1681	3394.80
30	73.50	900	2205.00
28	59.40	784	1663.20
11	25.60	121	281.60
64	133.86	4096	8567.04
57	148.80	3249	8481.60
46	112.20	2116	5161.20
50	101.20	2500	5060.00
48	116.60	2304	5596.80
33	83.60	1089	2758.80
46	112.20	2116	5161.20
64	143.52	4096	9185.28
13	28.80	169	374.40
54	108.56	2916	5862.24
70	120.00	4900	8400.00
49	118.80	2401	5821.20
28	72.60	784	2032.80
14	41.80	196	585.20
27	51.20	729	1382.40
46	122.40	2116	5630.40
22	64.80	484	1425.60
39	79.20	1521	3088.80

Best-fit linear equation is y = 2.1221x + 5.42503

Table 13-5.   Sales of new product, by month.

Month	Sales ($000)
1	4.8
2	7.6
3	7.8
4	11.1
5	12.7
6	12.2
7	13.1
8	14.0
9	14.3

way, giving the best-fit equation for data in an interval. We will demonstrate the method here with a time series. Later in the chapter we will use a computer program to demonstrate moving regression with another type of series.

You use the same formulas to calculate m and b. The only change is that n will now be a selected number instead of being the total number of data points. Every time we include a new data point, we will discard the oldest. For example, when you add a new x to the sum of xs, you will subtract the oldest x value from the sum.

Example 13.3

Table 13-5 shows sales, in thousands, of a new product. As the month-to-month differences are constantly changing, it would be foolish to consider the first month's sales as input when forecasting sales for the tenth month. Let us use a moving interval of five months, and organize the calculations in a table.

The first calculation will give us an equation that is the best fit for months one through five. This calculation is shown in Table 13-6. Then you change the figures to cover months two through six, then three through seven, and so on. The last interval, months five through nine, will be the best for predicting sales in the tenth month.

Table 13-6.   Calculations for first interval of five.

x	y	x^2	xy
1	4.8	1.0	4.8
2	7.7	4.0	15.4
3	7.8	9.0	23.4
4	11.1	16.0	44.4
5	12.7	25.0	63.5
15	44.1	55.0	151.5

Now m and b can be calculated by substituting these sums into the formulas.

$$m = \frac{5(151.5) - 15(44.1)}{5(55.0) - 15^2}$$

$$= 1.92$$

$$b = \frac{44.1 - 1.92(15)}{5}$$

$$= 3.06$$

The equation of the line that best fits the first five months is

$$\text{Sales} = 1.92(\text{month}) + 3.06$$

Instead of doing an entire set of new calculations for months two through six, you can just change the sums by subtracting the data from the first month and adding data from the sixth month.

$$\Sigma x = 15 - 1 + 6$$

$$= 20$$

$$\Sigma y = 44.1 - 4.8 + 12.2$$

$$= 51.5$$

$$\Sigma x^2 = 55.0 - 1.0 + 36.0$$

$$= 90.0$$

$$\Sigma xy = 151.5 - 4.8 + 73.2$$

$$= 219.9$$

Then m and b can be found by substituting these sums into the formulas. They are 1.39 and 4.74 respectively. The equation that best fits the data for months two through six is

$$\text{sales} = 1.39(\text{month}) + 4.74$$

You could continue in that manner, adding a new month to the sums and dropping the oldest month, keeping an interval of five months. However, if your only interest is the last five months, you do not have to work up to that interval; you can go right to that calculation, as shown in Table 13-7.

Substituting the sums in Table 13-7 into the formulas gives $m = 0.59$ and $b = 9.31$. Then the regression equation is

$$\text{sales} = 0.59(\text{month}) + 9.31$$

Table 13-7. Calculations for months five through nine.

x	y	x^2	xy
5	12.7	25.0	63.5
6	12.2	36.0	73.2
7	13.1	49.0	91.7
8	14.9	64.0	119.2
9	14.3	81.0	128.7
35	67.2	255.0	476.3

as the line that best fits the data of the last five months. That equation can be used to predict sales in the tenth month.

$$\text{sales in tenth month} = 0.59(10) + 9.31$$
$$= 15.21$$

The month-to-month change in sales is getting smaller. Considering this flattening (leveling) trend in the sales, this forecast seems much more reasonable than one based on the entire nine-month sales history.

As with moving average, the shorter the interval, the more the forecast will reflect the latest results. A long interval has a smoothing effect but it takes longer before a permanent change is reflected in the regression equation. The chapter on forecasting suggests ways to smooth forecasts so they don't respond too strongly to sudden short-lived changes.

COMPUTER PROGRAM FOR CALCULATING REGRESSION EQUATIONS

The amount of arithmetic in regression calculations makes them tedious and makes errors likely. Approaching the work systematically with a table reduces the chance of error but, even with a calculator, it is a long job. The computer program listed in Figure 13-1 does all the calculations and prints the best-fit equation. It even makes a prediction based on the equation. All that is necessary is that you enter the data points.

How the Program Works

The program uses standard BASIC language commands and should work on most computers with few if any changes.

Nine choices of action

The menu displayed by lines 130 through 260 offers nine choices of action. The first choice must be 7, 8, or 9 in order to place data in the computer's memory. Then you can choose to see either the entire table of calculations or just the best-fit equation. You can choose to have the calculations include all the data or a selected interval.

Example 13.4

Table 13-8 is a copy of the records showing, for fifty employees, the number of weeks they have been assembling the current product, and the number they assembled last week. From this data, predict how many units a worker with forty weeks experience will assemble next week.

Run the program and select number seven from the menu. The screen will display the message

Enter *x* value #1 . . . or END

Type the first employee's number of weeks (16) and press **ENTER**. The screen

Figure 13-1. Computer program for regression calculations.

```
10 '*********************************************************************
20 '** REGLIN.BAS  Linear regression for either all data or a moving   **
30 '** interval.   For Handbook of Manufacturing and Production        **
40 '** Management Formulas, Charts, and Tables                         **
50 '*********************************************************************
60 U1$="#####.####":U2$="#######0######"
70 DIM X(300),Y(300)
80 CLS
90 PRINT
100 PRINT"       Linear regression with all data"
110 PRINT"  1. Print just best-fit equation on screen"
120 PRINT"  2. Print table of all calculations on screen"
130 PRINT"  3. Print table on screen and line printer"
140 PRINT
150 PRINT"       Moving linear regression"
160 PRINT"  4. Print just best-fit equation on screen"
170 PRINT"  5. Print table of all calculations on screen"
180 PRINT"  6. Print table on screen and line printer"
190 PRINT
200 PRINT"  7. Enter data from keyboard"
210 PRINT"  8. Retrieve data from disk"
220 PRINT"  9. Save data on disk"
230 INPUT"Select by number ",S
240 IF S>0 AND S<10 THEN 270
250 PRINT"select 1 through 9 only please!!!"
260 GOTO 90
270 CLS
280 ON S GOTO 300,330,360,390,420,450,480,1010,1100
290 '(1) All data, give just equation  111111111
300 SP=0:LP=0
310 GOTO 580
320 '(2) All data, table on screen  222222222
330 SP=1:LP=0
340 GOTO 580
350 '(3) All data, table on screen & printer  3333333
360 SP=1:LP=0
370 GOTO 580
380 '(4) Moving interval, give just equation  44444444
390 SP=0:LP=0
400 GOTO 710
410 '(5) Moving interval, table on screen  55555555
420 SP=1:LP=0
430 GOTO 710
440 '(6) Moving interval, table on screen and printer  66666666
450 SP=1:LP=1
460 GOTO 710
470 '(7) Enter data from keyboard  7777777
480 N=0
490 PRINT"Enter X value #";N+1;"... OR END";
500 INPUT X$
510 IF X$>="E" THEN 90
520 N=N+1
530 X(N)=VAL(X$)
540 INPUT"      Y value ",Y(N)
550 PRINT
560 GOTO 490
570 '(1),(2),(3) Linear regression with all data  1 2 3 1 2 3 1 2 3
580 GOSUB 1180           'Reset accumulators
590 NT=N
600 IF SP=1 THEN GOSUB 1200       'Table headings on screen
610 IF LP=1 THEN GOSUB 1230       'Table headings on printer
620 FOR C=1 TO N
630   GOSUB 1260                  'Main calculations & table
```

Figure 13-1. (continued)

```
640 NEXT C
650 GOSUB 1390                    'Finish equation
660 GOSUB 1520                    'Request next step
670 ON T GOTO 680,490,80
680 GOSUB 1600                    'Make calculation
690 GOTO 660
700 '(4),(5),(6) Moving linear regression  4 5 6 4 5 6
710 INPUT"Enter number of lowest data point to be considered ",PL
720 INPUT"... and the highest ",PH
730 IF PH<=N THEN 770
740 PRINT"There are only";N;"data points!!!"
750 PRINT:INPUT"Enter number of highest data point to be considered ",PH
760 GOTO 730
770 IF PH>PL THEN 800
780 PRINT"Highest must be higher than lowest!!!"
790 PRINT:GOTO 710
800 INPUT"How many data points in the moving interval? ",I
810 IF I<PH-PL+2 THEN 840
820 PRINT"There aren't that many data points between the"
830 PRINT"lowest and highest you selected!!!":GOTO 710
840 L=PL
850 GOSUB 1180                    'Reset accumulators
860 NT=I
870 IF SP=1 THEN GOSUB 1200       'Table headings on screen
880 IF LP=1 THEN GOSUB 1230       'Table headings on printer
890 FOR C=L TO L+I-1
900    GOSUB 1260                 'Main calculations & table
910 NEXT C
920 GOSUB 1390                    'Finish equation
930 GOSUB 1520                    'Request next step
940 ON T GOTO 950,970,80
950 GOSUB 1600                    'Make calculation
960 GOTO 930
970 IF L<PH-I+1 THEN 990
980 PRINT"All intervals completed":GOTO 90
990 L=L+1:PRINT:GOTO 850
1000 '(8) Retrieve data from disk  8888888
1010 LINE INPUT"Name of file on disk ",N$
1020 OPEN"I",1,N$
1030 N=0
1040 IF EOF(1) THEN 1080
1050 N=N+1
1060 INPUT#1,X(N),Y(N)
1070 GOTO 1040
1080 CLOSE:PRINT"Data retrieval complete.":GOTO 90
1090 '(9) Save data on disk  99999999
1100 LINE INPUT"Name for saving data ",N$
1110 OPEN"O",1,N$
1120 FOR C=1 TO N
1130    WRITE#1,X(C),Y(C)
1140 NEXT C
1150 CLOSE:PRINT"Data storage complete.":GOTO 90
1160 END
1170 'Subroutine to reset accumulators to zero
1180 SB=0:SX=0:SY=0:SS=0:RETURN
1190 'Subroutine to print table headings on screen
1200 PRINT TAB(7)"x";TAB(22)"y";TAB(36)"x sq";TAB(54)"xy"
1210 RETURN
1220 'Subroutine to print table headings on printer
1230 LPRINT TAB(7)"x";TAB(22)"y";TAB(36)"x sq";TAB(54)"xy"
1240 LPRINT STRING$(62,"."):RETURN
1250 'Subroutine for main calculations
1260 XY=X(C)*Y(C):SB=SB+XY:SX=SX+X(C):SY=SY+Y(C):XS=X(C)*X(C):SS=SS+XS
```

Figure 13-1. (continued)

```
1270 IF SP=0 THEN 1320
1280 PRINT TAB(2)USING U1$;X(C);
1290 PRINT TAB(16)USING U1$;Y(C);
1300 PRINT TAB(30)USING U2$;XS;
1310 PRINT TAB(46)USING U2$;XY
1320 IF LP=0 THEN 1370
1330 LPRINT TAB(2)USING U1$;X(C);
1340 LPRINT TAB(16)USING U1$;Y(C);
1350 LPRINT TAB(30)USING U2$;XS;
1360 LPRINT TAB(46)USING U2$;XY
1370 RETURN
1380 'Subroutine to finish equation
1390 M=(NT*SB-SX*SY)/(NT*SS-SX*SX)
1400 B=(SY-M*SX)/NT
1410 IF B<0 THEN SG$="-" ELSE SG$="+"
1420 PRINT
1430 IF S<>4 THEN 1450
1440 PRINT"Data points";L;"through";L+I-1
1450 PRINT"Best-fit linear equation is  y=";M;"x ";SG$;ABS(B)
1460 PRINT
1470 IF LP=0 THEN 1500
1480 LPRINT:LPRINT"Best-fit linear equation is y=";M;"x ";SG$;ABS(B)
1490 LPRINT
1500 RETURN
1510 'Subroutine to request next step
1520 PRINT"Type 1 to make a calculation with this equation"
1530 PRINT"     2 to continue"
1540 PRINT"     3 to return to main selections"
1550 INPUT T
1560 IF T>0 AND T<4 THEN 1580
1570 PRINT"Type 1, 2, or 3 please!!!":PRINT:GOTO 1520
1580 CLS:RETURN
1590 'Subroutine to make calculation with equation
1600 INPUT"Enter x value for calculation ",X
1610 PRINT"y(";X;")=";M*X+B:PRINT
1620 IF LP=0 THEN 1640
1630 LPRINT"y(";X;")=";M*X+B:LPRINT
1640 RETURN
```

will then ask for the y value; type the units the first worker assembled (52) and press **ENTER**. Then the screen will ask for data from 2, and so forth. When it asks for x value number 51, type **END** and press **ENTER.**

Line 540 then sends the program back to display the menu again, and you can choose the type of display. For this example, choose number one; the computer will quickly make the calculations and then display

Best-fit linear equation is $y = 1.50737x + 18.1461$

and it will ask if you want a calculation. For example, if you want to make a prediction for a worker with twenty weeks experience, type **1** and press **ENTER.** When the screen asks for an X value, type **20** and press **ENTER.** The screen will immediately show

$$y(20) = 46.2935$$

which is the prediction, based on this group of fifty employees.

The program allows you to continue making predictions as long as you type

1 in response to its choices. When you type **3**, the menu returns and you can choose the next action. If you choose **1** through **6**, the computer will use the same data you entered into its memory. Therefore, you can choose to see just the equation as done in this example, have the entire table printed, and then look at the results for some interval within the data. As long as you do not choose **7** or **8**, the data in memory remains unchanged and you can have a variety of calculations performed on it.

Example 13.5

In this example you will look at regression equations in segments by adapting the moving interval feature of the computer program. Production machines are used from zero to sixty hours per week, depending on orders received the week before. Table 13-9 shows, for various weeks in the past, the rate at which machines break down with different amounts of use.

Rate of breakdown

A glance at the figures reveals that the *rate* of breakdown increases with weekly usage. In fact, for high weekly usages, the rate can be seen to increase ever more rapidly, indicating that the graph curves. Therefore, you will have a more realistic predictor of failure rate if you derive different regression equations for different segments of the usage scale.

First rearrange the data with the least number of hours first, as shown in Table 13-10. Then you can examine the data to see if the values form a pattern.

It appears from Table 13-10 that the pattern changes with data value number 18, and again with number 38. You can therefore obtain a regression equation for each of the groups.

Run the regression computer program and it will display a menu of nine items. Select number **7** because you will first enter the fifty data items from Table 13-10 via the keyboard. Then type **END** and the menu will be displayed again. This time select number **4** because you can use the moving interval feature to get the equation for a portion of the data values.

First interval

Begin with the first segment, which should consist of data items number 1 through 17. When the screen asks for the lowest data item number, type **1**. The highest will be 17, and there will be 17 items in the interval. Notice that you do not actually have a moving interval, but that part of the program works for this calculation if you let the interval consist of all the data items in which you are interested.

The computer then calculates the regression equation that fits data items 1 through 17 and prints

$$y = 4.76624 \, E\text{-}03x - .0245127$$

where E-03 means move the decimal point three places to the left. Written in a more familiar form, the equation is

$$y = .00476624x - .0245127$$

Table 13-8. Productivity as a function of experience.

	Hours	Assemblies
1	16.00	52.00
2	6.00	27.00
3	13.00	29.00
4	12.00	42.00
5	20.00	47.00
6	13.00	40.00
7	18.00	51.00
8	20.00	54.00
9	12.00	38.00
10	15.00	40.00
11	17.00	38.00
12	6.00	22.00
13	20.00	49.00
14	19.00	48.00
15	20.00	40.00
16	10.00	25.00
17	15.00	43.00
18	2.00	16.00
19	11.00	35.00
20	19.00	48.00
21	13.00	29.00
22	8.00	25.00
23	12.00	38.00
24	13.00	32.00
25	22.00	42.00
26	25.00	55.00
27	17.00	45.00
28	13.00	32.00
29	20.00	40.00
30	12.00	35.00
31	14.00	41.00
32	12.00	36.00
33	4.00	21.00
34	19.00	35.00
35	7.00	24.00
36	19.00	45.00
37	13.00	37.00
38	16.00	36.00
39	21.00	55.00
40	15.00	36.00
41	6.00	29.00
42	17.00	37.00
43	25.00	47.00
44	22.00	52.00
45	11.00	27.00
46	4.00	24.00
47	6.00	29.00
48	8.00	34.00
49	25.00	53.00
50	2.00	15.00

Table 13-9. Machine breakdown rate.

	Hours	Breakdowns
1	7.00	0.01
2	58.00	0.81
3	49.00	0.53
4	50.00	0.47
5	48.00	0.47
6	31.00	0.15
7	9.00	0.02
8	15.00	0.05
9	34.00	0.18
10	37.00	0.33
11	57.00	0.58
12	13.00	0.03
13	54.00	0.70
14	60.00	0.65
15	11.00	0.03
16	52.00	0.43
17	9.00	0.02
18	9.00	0.02
19	27.00	0.15
20	5.00	0.01
21	20.00	0.10
22	18.00	0.06
23	13.00	0.03
24	25.00	0.14
25	14.00	0.04
26	26.00	0.16
27	38.00	0.32
28	16.00	0.06
29	23.00	0.12
30	16.00	0.05
31	37.00	0.30
32	42.00	0.42
33	7.00	0.01
34	17.00	0.06
35	50.00	0.45
36	37.00	0.28
37	27.00	0.13
38	44.00	0.31
39	43.00	0.33
40	35.00	0.29
41	11.00	0.03
42	34.00	0.16
43	26.00	0.15
44	59.00	0.77
45	50.00	0.40
46	33.00	0.17
47	53.00	0.67
48	54.00	0.70
49	44.00	0.35
50	8.00	0.02

Table 13-10. Data arranged in ascending order.

	Hours	Breakdowns
1	5.00	0.01
2	7.00	0.01
3	7.00	0.01
4	8.00	0.02
5	9.00	0.02
6	9.00	0.02
7	9.00	0.02
8	11.00	0.03
9	11.00	0.03
10	13.00	0.03
11	13.00	0.03
12	14.00	0.04
13	15.00	0.05
14	16.00	0.05
15	16.00	0.06
16	17.00	0.06
17	18.00	0.06
18	20.00	0.10
19	23.00	0.12
20	25.00	0.14
21	26.00	0.15
22	26.00	0.16
23	27.00	0.13
24	27.00	0.15
25	31.00	0.15
26	33.00	0.17
27	34.00	0.16
28	34.00	0.18
29	35.00	0.29
30	37.00	0.30
31	37.00	0.28
32	37.00	0.33
33	38.00	0.32
34	42.00	0.42
35	43.00	0.33
36	44.00	0.35
37	44.00	0.31
38	48.00	0.47
39	49.00	0.53
40	50.00	0.47
41	50.00	0.45
42	50.00	0.40
43	52.00	0.43
44	53.00	0.67
45	54.00	0.70
46	54.00	0.70
47	57.00	0.58
48	58.00	0.81
49	59.00	0.77
50	60.00	0.65

The computer's answer contains more decimal places than justified by the input data. Whenever the machines are used eighteen hours per week or less, you can predict the breakdown rate from

breakdown rate = .0048(hours used in week) − .0245

or you could round it still further.

Second interval

For the next interval, return to the menu and select **4** again. This time take data items number 18 through 37, which is twenty items. The regression equation for this interval is

$$y = .0120272x − .172052$$

which you should round to

breakdown rate = .012(hours used per week) + .172

When weekly usage is between in this interval, this equation will predict the failure rate we can generally expect.

Third interval

The third interval covers thirteen data values, numbers 38 through 50. The computer prints the equation

$$y = .0262709x − .816232$$

which should be rounded for predicting breakdown rates in this interval.

breakdown rate = .0263(hours used in week) − .816

By dividing the weekly usage rate into three intervals you have derived equations that predict breakdown rates more realistically than one overall equation would. However it was necessary that the original time series (Table 13-9) be rearranged in ascending order of weekly usage rates.

NONLINEAR REGRESSION

When the data form a curve and it is necessary to find a single equation that is the best fit, there are various equations other than a straight line that can be found. However, even statistics books mark this topic optional — for advanced students. Nonlinear regression is not included here; very good explanations can be found in books such as John E. Freund, *Modern Elementary Statistics*, fourth edition, 1973; Prentice-Hall, Inc., Englewood Cliffs, New Jersey.

CHAPTER 14

Simulation: A Systematic Approach to Testing Decisions

This chapter explores the process of "simulation," whereby real-life situations are studied for their projected effects on a proposed course of action or policy. Management training programs often employ simulation "games" as part of the learning procedure.

The simulation process parallels actual conditions by arranging for random numbers to select events with the same probability as the actual events they represent. For example, suppose you know that $\frac{1}{6}$ of the orders you receive are for product Q, and $\frac{5}{6}$ are for product R. You can simulate incoming orders by rolling a die; a 1, 2, 3, 4, or 5 indicates an order for product R and a 6 indicates an order for product Q.

Most simulations are run for one of two reasons. First, it is less expensive than finding the answers through real life. For example, you might want to know how different sequences in a production process (shear, then punch, then stamp versus punch, then stamp, then shear) will affect the amount of bin storage required. Instead of trying each sequence for a few months, you can simulate them at much less cost.

The other reason is that some situations cannot be repeated with different arrangements. You might have to decide whether to send winter shipments of a twelve-month contract by barge, rail, or truck. If you could try each method for one winter, you would know by the fourth winter which is best. As you do not have that much time, you can simulate a winter with each method — with a computer you can simulate hundreds of winters.

You use a model to simulate operation of the actual system, but a model could never be as complex as a real-life system. The more the model acts like the system, the more reliable the results of the simulation will be. On the other hand, if the model is too complex, running it can be as complex as running the actual system. The major decision in conducting a simulation is selecting the level of complexity for the model. That is, finding the point of diminishing returns, where increasing complexity does not add significantly to the model's realism.

We know the long-run rule, but the short-run details that accumulate to the long-run are random. That is, although the next event is random, the cumulative effect of the random events is predictable.

CHOOSING RANDOM NUMBERS

For a series of numbers to be random it is necessary that, for the next number, each of the possibilities is equally likely, regardless of the past pattern. Because humans are not good at picking numbers randomly, many books give tables (see Table A-2 in the appendix). Most computers can provide seemingly endless series of random numbers.

Although each number is random, you shape the overall results by grouping random numbers. For example, suppose you want to simulate an assembly line in which everyone is present $\frac{1}{10}$ of the time, one person is absent half the time, two are absent $\frac{3}{10}$ of the time, and three are absent $\frac{1}{10}$ of the time. You can select digits zero through nine randomly, assigning them as shown in Table 14-1.

Table 14-1. Random numbers assigned so event simulated occurs with same probability as actual event.

Random digits	Number absent
0	0
1–5	1
6–8	2
9	3

Now you can take digits from a random number table and, *in the long-run*, perfect attendance will be indicated ten percent of the time, one absence fifty percent of the time, and so on.

When the random numbers needed do not use all the numbers in a series (for example, all one-digit numbers, or all two-digit numbers), there are two ways to proceed. Suppose, for example, you are using only three numbers. You could assign 0 to one event, 1 to another, and 2 to another; then whenever 3 through 9 appeared you would disregard them and draw another number until drawing a 0, 1, or 2. Of course, you might spend a lot of time drawing numbers and disregarding them, and you might also exhaust the random number table.

Another method is to assign random numbers so as to use as many as possible of the digits. You can say that either a 0, a 1, or a 2 represents one event; a 3, 4, or 5 the second event, and a 6, 7, or 8 the third event. Then you have to disregard and redraw only for a 9.

MANUAL SIMULATION: FOR LESS COMPLICATED EVENTS

If the event is not complicated, and only a few cycles of the event are to be simulated, the entire process can be done with pencil, paper, and a random number table. Let's look at one such example, followed by some computer simulations.

Example 14.1

A new manager is told that the production department has six employees assigned to run the machines for a certain product. The practice has always been that they are randomly assigned in two groups of three each day, and each group is assigned to one of three machines for the day. One machine is idle each day.

When workers A, B, and C make up a team they produce twenty good units and one reject for the day. If a team is made of workers A, C, and D, they produce twelve good units and four rejects. Any other combination produces eighteen good units and two rejects.

Any rejects from machine one must be scrapped, regardless of which team used it. Rejects from machines two and three can be reworked, the firm considers those rejects as a half unit.

Not wanting to continue a policy just because "it has always been done that way," the new manager simulates the arrangement, as well as other arrangements to see what output is likely from each arrangement.

A little experimenting with letters A to F shows that, besides ABC and ACD, eight other groups can be formed. (See also "Combinations" in Chapter 12.) Therefore the manager draws random digits zero through nine, deciding that team ABC is formed each time zero is drawn, team ACD each time one is drawn, and an "other" team for two through nine.

Assigning machines cannot be done as easily. There is a $\frac{1}{3}$ probability of any machine's being assigned to any given group. Then, having assigned one machine, there is a $\frac{1}{2}$ probability for each of the other machine's being assigned to the other group. Step 1 of the machine assignment is to draw random numbers 0 through 9, let 0 to 2 assign machine 1, 3 to 5 assign machine 2, 6 to 8 assign machine 3, and disregard each 9 that is drawn. Then draw another number 0 through 9, letting 0 through 4 assign the lower of the remaining two machines (1 if either machine 2 or 3 was assigned first; 2 if machine 1 was assigned first).

The final step before running the simulation is to prepare a table to be filled in. Table 14-2 shows one possible arrangement for the data.

Let's use Table A-2 as a source of random numbers. For teams, use the digits in the fifth group of five on each row; the first machine assignment will be from the second group of five, and the second machine assignment from the third group of five. Table 14-3 shows the table filled in for ten days.

Notice that on the ninth day you drew 9 for the first machine assignment. However, you had already established the rule that any 9 would be disregarded; therefore, you used 1, which came up next.

Now the manager can simulate other arrangements, such as keeping the best team together on machine 2 or 3. After simulating other possible arrangements, he or she will know what to expect when establishing a new policy.

COMPUTER SIMULATION: FOR LONG OR REPEATED SIMULATIONS

Because simulation is based on randomness, you do not know what the outcome might be if you ran the simulation again. Neither do you know that the real-life outcome will be the same as the simulated outcome. The simulation shows only a typical, or likely, outcome.

Advantage of Computer Simulation

However, there is a very big advantage to the method of simulation — it is cheap and easy to run a large number of times. It is not advisable for a manager

Table 14-2.   Headings for table to record results of manual simulation.

Day	Rnd nbr	Team	Produced good	rej	Rnd nbr	Mach	Team net	Day's net

Table 14-3. Table filled in to simulate ten days.

Day	Rnd nbr	Team	Produced good	rej	Rnd nbr	Mach	Team net	Day's net
1	5	other	18	2	1	1	18	
		other	18	2	2	2	19	37
2	5	other	18	2	3	2	19	
		other	18	2	3	1	18	37
3	1	ACD	12	4	5	2	14	
		other	18	2	2	1	18	32
4	3	other	18	2	1	1	18	
		other	18	2	8	3	19	37
5	8	other	18	2	4	2	19	
		other	18	2	7	3	19	38
6	2	other	18	2	6	3	19	
		other	18	2	2	1	18	37
7	0	ABC	20	1	5	2	20.5	
		other	18	2	0	1	19	39.5
8	3	other	18	2	8	3	19	
		other	18	2	7	2	19	38
9	5	other	18	2	9 → 1	1	18	
		other	18	2	0	2	19	37
10	0	ABC	20	1	6	3	20.5	
		other	18	2	8	2	19	39.5

to make a decision after simulating only ten day's production. It is not advisable even after observing the *actual* result of only ten day's production. But it is usually not practical to observe 1,000 day's production before making a decision, whereas it is easy to simulate 1,000 day's production. In addition, the manager can simulate 1,000 day's production with each of the policies under consideration.

If you made more runs of ten days, you should not be surprised if the results were quite different from each other. However, in 1,000 days the randomness smooths out and you an expect the results of successive runs to be more nearly alike. We can also expect that a long simulation will more nearly indicate the results of a long actual run.

Although it is possible to continue the simulation of Table 14-3 to 1,000 days, it is better to program a computer to do the work. The computer can be programmed to print each day's results, as in Table 14-3, or it can just accumulate them and print the final result.

Procedure for Using Computer Simulation

It is not difficult to program a computer to follow the same procedure followed in the manual example. Figure 14-1 does the following:

1. pick the next day number
2. pick the first random number

Figure 14-1.   Computer program that does simulation in Table 14-3.

```
10 '******************************************************************
20 '** TEAMSIM.BAS  Computer version of simulation in        **
30 '** Table 14-3.  Handbook of Manufacturing and Production  **
40 '** Management Formulas, Charts, and Tables                **
50 '******************************************************************
60 RANDOMIZE
100 CLS
110 SCREEN 0
120 WIDTH 80
130 DAYFMT$="##"
140 DAY=0
150 INPUT "How many days to be simulated? ",LASTDAY
160 CLS
170 PRINT TAB(7)"Random";TAB(23)"Production";TAB(35)"Random";TAB(57)"Net";TAB(67
)"Daily"
180 PRINT "Day";TAB(7)"number";TAB(16)"Team";TAB(23)"good";TAB(29)"rej";TAB(35)"
number";TAB(44)"Machine";TAB(54)"production";TAB(68)"net"
190 '(1) Start next day  <<<<<<<<<<<<<
200 DAY=DAY+1
210 GOSUB 1020
220 GOSUB 1210
230 GOSUB 1320
240 GOSUB 1410
250 GOSUB 1510
260 GOSUB 1610
270 GOSUB 1810
280 GOSUB 1910
290 IF DAY>=LASTDAY THEN STOP
300 GOTO 200
310 END
1000 '(2) & (3) Composition of first team  $$$$$$$$$$$$$$$
1010 'Pick random number  <<<<<<<<<<<<<
1020 RN1=INT(RND*10)
1030 'Identify team  <<<<<<<<<<<<
1040 IF RN1=0 THEN TEAM$="ABC": GOTO 1070
1050 IF RN1=1 THEN TEAM$="ACD": GOTO 1070
1060 TEAM$="other"
1070 RETURN
1200 '(4) Determine production of first team  $$$$$$$$$$$$$$$
1210 IF TEAM$="ABC" THEN GOOD=20: REJ=1: GOTO 1240
1220 IF TEAM$="ACD" THEN GOOD=12: REJ=4: GOTO 1240
1230 GOOD=18: REJ=2
1240 RETURN
1300 '(5) & (6) Identify machine for 1st team  $$$$$$$$$$$$
1310 'Pick second random number  <<<<<<<<<<<<<
1320 RN2=INT(RND*10)
1330 'Relate random number to machine number  <<<<<<<<<<<<<<
1340 IF RN2<3 THEN MACHINE1=1: GOTO 1380
1350 IF RN2<6 THEN MACHINE1=2: GOTO 1380
1360 IF RN2<9 THEN MACHINE1=3: GOTO 1380
1370 GOTO 1320          'Draw another if number was 9
1380 RETURN
1400 '(7) Determine net production of 1st team  $$$$$$$$$$$$
1410 IF MACHINE1=1 THEN NET=GOOD: GOTO 1430
1420 NET=GOOD+REJ/2
1430 DAILYNET=NET
1440 RETURN
1500 '(8) Print first line of day  $$$$$$$$$$$$$$$
1510 PRINT TAB(2) USING DAYFMT$; DAY;
1520 PRINT TAB(9) RN1;TAB(16)TEAM$;TAB(23)GOOD;TAB(29)REJ;TAB(37)RN2;TAB(46)MACH
INE1;TAB(57)NET
1530 RETURN
1600 '(9) & (10) Identify machine for 2nd team  $$$$$$$$$$$$
```

Figure 14-1. (continued)

```
1610 IF MACHINE1=1 THEN M2=2: M3=3: GOTO 1640
1620 IF MACHINE1=2 THEN M2=1: M3=3: GOTO 1640
1630 M2=1: M3=2
1640 RN3=INT(RND*10)
1650 IF RN3<5 THEN MACHINE2=M2: GOTO 1670
1660 MACHINE2=M3
1670 RETURN
1800 '(11) & (12) Net production of 2nd team & total for day  $$$$$$$$$$
1810 IF MACHINE2=1 THEN NET=18: GOTO 1830
1820 NET=19
1830 DAILYNET=DAILYNET+NET
1840 RETURN
1900 '(13) Print second line for day  $$$$$$$$$$$$$$$
1910 PRINT TAB(16)"other";TAB(23)18;TAB(29)2;TAB(37)RN3;TAB(46)MACHINE2;TAB(57)N
ET;TAB(68)DAILYNET
1920 PRINT
1930 RETURN
```

3. identify the composition of a team, based on a predetermined rule for applying random numbers

4. determine how many good and how many rejected units were produced, according to which team is working

5. pick the second random number

6. determine which machine that team worked on, based on a predetermined rule for applying random numbers

7. determine the net production, according to which machine the team worked on

8. display a line of data for Team 1

9. pick the third random number

10. determine which machine the second team (it will an an "other" team) worked on, based on a predetermined rule for applying random numbers

11. determine the net production, according to which machine the team worked on

12. calculate the daily effective production

13. display second line of data for day

14. if more days are to be simulated, repeat all steps

Step number ten states that one team each day will always have an "other" composition. That is because employees A and C are in both of the identified teams; there cannot be both an ABC and an ACD team the same day.

The computer program listed in Figure 14-1 performs each of these steps. Each step is identified in a remark, so you can easily determine which part of the program does each step. This program was written for and tested on IBM PC and PCjr, but standard BASIC was used so it will run on other computers that use BASIC.

RANDOMIZE in line 60 means the computer will ask you (when you run the program) for a "kernel" number to start the random number generator. Enter any number in the range −32768 to 32767; the computer will generate a different set of psuedo-random numbers for every different number you enter. If you want to repeat a simulation exactly, enter the same kernel. This is the line that is most likely to have to be tailored for the specific computer used.

Table 14-4 shows the program used for a sixteen-day simulation.

HOW TO WRITE A COMPUTER PROGRAM FOR SIMULATIONS THAT ARE MORE INVOLVED: THREE EXAMPLES

The next examples demonstrate simulations of medium complexity — using the computer programs is more desirable than manual operations. As every simula-

Table 14-4. Computer simulation Example 14.1 for sixteen days.

7 Day	Random number	Team	Production good	rej	Random number	Machine	Net production	Daily net
1	8	other	18	2	8	3	19	
		other	18	2	8	2	19	38
2	1	ACD	12	4	5	2	14	
		other	18	2	0	1	18	32
3	6	other	18	2	6	3	19	
		other	18	2	1	1	18	37
4	6	other	18	2	2	1	18	
		other	18	2	8	3	19	37
5	7	other	18	2	0	1	18	
		other	18	2	2	2	19	37
6	0	ABC	20	1	4	2	20.5	
		other	18	2	2	1	18	38.5
7	0	ABC	20	1	0	1	20	
		other	18	2	7	3	19	39
8	4	other	18	2	5	2	19	
		other	18	2	5	3	19	38
9	3	other	18	2	0	1	18	
		other	18	2	0	2	19	37
10	5	other	18	2	6	3	19	
		other	18	2	3	1	18	37
11	3	other	18	2	7	3	19	
		other	18	2	2	1	18	37
12	6	other	18	2	6	3	19	
		other	18	2	9	2	19	38
13	8	other	18	2	2	1	18	
		other	18	2	0	2	19	37
14	0	ABC	20	1	8	3	20.5	
		other	18	2	5	2	19	39.5
15	5	other	18	2	8	3	19	
		other	18	2	8	2	19	38
16	7	other	18	2	2	1	18	
		other	18	2	0	2	19	37

tion is unique, it is not likely you will be able to use any of them directly; they are meant only to demonstrate how to write a computer program once you have determined the details of your simulation.

Example I: Manufacturing for Inventory Below a Trigger Level

This program simulates a factory that begins manufacturing whenever their inventory gets below a trigger level. The output goes into inventory, while orders continue to arrive and items are removed from inventory to ship. If the manufacturing rate is faster than the order rate, inventory will build up; at a certain level of inventory they suspend manufacturing until inventory returns to the trigger level.

The situation is further complicated because manufacturing is done by a team that is dispersed to other areas of the company during suspension of production. Therefore, it takes a certain number of days from the time inventory reaches its trigger level until production actually begins.

This type of simulation would be used if management wanted to keep inventory costs low while maintaining a reputation for "same day shipment." They could experiment with the trigger level and the stop-producing level because those variables are certainly under their control. The number produced per day could also be changed, but not as easily. Start-up time is shown as a variable, but might be the hardest to change and might not be considered under the manager's control.

Table 14-5 shows a simulation of fifty days. The next to the last column, backlog, shows how often the firm was unable to ship material the day the order was received. The policy assumed for writing this program was that the company shipped as much as it could; if orders for ten were received and there were two in stock, then the two were shipped and eight were listed as backlog.

The program is listed in Figure 14-2. For longer simulations, it might be better to change the program so the computer simply counts the number of days that ended with a backlog, instead of printing the results of each day. Comments by many of the lines help you determine the purpose of each area of the program.

Example II: Automatic Machine Setting

In Chapter 2 you saw that setting a machine to cut to the center of a tolerance band is not always the most cost-effective setting. For example, suppose that when the machined part is too large it can be machined further, but when it is too small it must be scrapped. As it costs more when the machined size is too small, the target setting should be on the high side of center.

Table 2-3 gave the results of twenty-one simulations, each simulating the manufacture of 5,000 pieces. The table, clearly showing the least-cost setting, was made by the program that is listed in Figure 14-3.

Example III: Sharpening and Replacing a Tool

The last simulation demonstrated examines a familiar problem: if a tool is sharpened frequently, it makes the product more reliable, but sharpening costs

Table 14-5. Simulation of manufacturing for inventory.

Run
How many days are to be simulated? 50
At what inventory level do we trigger manufacturing? 20
At what inventory level do we stop manufacturing? 35
How many units do we manufacture per day? 10
How many days startup time required? 4

Day	Mfg. Triggered	Producing	Orders Rcvd	Shipped for Backlog	Shipped for New order	Ending Backlog	Ending Inventory
1	no	no	7	0	7	0	28
2	no	no	7	0	7	0	21
3	no	no	6	0	6	0	15
4	yes	no	14	0	14	0	1
5	yes	no	7	0	1	6	0
6	yes	no	3	0	0	9	0
7	yes	no	5	0	0	14	0
8	yes	yes	6	10	0	10	0
9	yes	yes	3	10	0	3	0
10	yes	yes	11	3	7	4	0
11	yes	yes	4	4	4	0	2
12	yes	yes	4	0	4	0	8
13	yes	yes	4	0	4	0	14
14	yes	yes	10	0	10	0	14
15	yes	yes	6	0	6	0	18
16	yes	yes	5	0	5	0	23
17	yes	yes	0	0	0	0	33
18	yes	yes	8	0	8	0	35
19	no	no	1	0	1	0	34
20	no	no	5	0	5	0	29
21	no	no	6	0	6	0	23
22	no	no	6	0	6	0	17
23	yes	no	3	0	3	0	14
24	yes	no	5	0	5	0	9
25	yes	no	8	0	8	0	1
26	yes	no	9	0	1	8	0
27	yes	yes	5	8	2	3	0
28	yes	yes	7	3	7	0	0
29	yes	yes	0	0	0	0	10
30	yes	yes	6	0	6	0	14
31	yes	yes	6	0	6	0	18
32	yes	yes	8	0	8	0	20
33	yes	yes	3	0	3	0	27
34	yes	yes	6	0	6	0	31
35	yes	yes	9	0	9	0	32
36	yes	yes	8	0	8	0	34
37	yes	yes	3	0	3	0	41
38	no	no	6	0	6	0	35
39	no	no	3	0	3	0	32
40	no	no	4	0	4	0	28
41	no	no	4	0	4	0	24
42	no	no	5	0	5	0	19
43	yes	no	10	0	10	0	9
44	yes	no	4	0	4	0	5
45	yes	no	6	0	5	1	0
46	yes	no	9	0	0	10	0
47	yes	yes	7	10	0	7	0
48	yes	yes	3	7	3	0	0
49	yes	yes	0	0	0	0	10
50	yes	yes	8	0	8	0	12

Enter C to continue; any letter to start over

Figure 14-2. Computer program simulating manufacturing for inventory.

```
10 '***********************************************************************
20 '** SIM.BAS  Simulation of factory orders for Handbook of Manufacturing **
30 '** and Production Management Formulas, Charts, and Tables            **
40 '***********************************************************************
50 KEY OFF
60 CLEAR,,,32768!
70 SCREEN 6
80 COLOR 3,9
90 DEFINT A-Z
100 U$="###"
110 DB=1
120 CLS
130 INPUT "How many days are to be simulated";DS
140 INPUT "At what inventory level do we trigger manufacturing";TM
150 INPUT "At what inventory level do we stop manufacturing";SM
160 INPUT "How many units do we manufacture per day";UD
170 INPUT "How many days startup time required";TR
180 I=SM: BL=0: M$="no": P$="no"
190 PRINT TAB(9)"Mfg.";TAB(28)"Orders";TAB(39)"Shipped for";TAB(61)"Ending"
200 PRINT "Day";TAB(6)"Triggered  Producing  Rcvd  Backlog  New order  Backlog
    Inventory"
210 FOR D=DB TO DB+DS-1
220   PRINT USING U$;D;
230   PRINT TAB(9) M$; TAB(19) P$;
240   SR=INT(RND*270)+1
250   IF SR<6 THEN S=0: GOTO 400
260   IF SR<14 THEN S=1: GOTO 400
270   IF SR<27 THEN S=2: GOTO 400
280   IF SR<53 THEN S=3: GOTO 400
290   IF SR<90 THEN S=4: GOTO 400
300   IF SR<130 THEN S=5: GOTO 400
310   IF SR<168 THEN S=6: GOTO 400
320   IF SR<200 THEN S=7: GOTO 400
330   IF SR<225 THEN S=8: GOTO 400
340   IF SR<242 THEN S=9: GOTO 400
350   IF SR<253 THEN S=10: GOTO 400
360   IF SR<261 THEN S=11: GOTO 400
370   IF SR<266 THEN S=12: GOTO 400
380   IF SR<269 THEN S=13: GOTO 400
390   S=14
400   IF P$="no" THEN 420
410   I=I+UD
420   IF BL<1 THEN SB=0: GOTO 520    'IF THERE IS NO BACKLOG
430   IF I>=BL THEN 490              'IF WE CAN FILL BACKLOG
440   SB=I                          'SHIP ALL WE HAVE TOWARD BACKLOG
450   SN=0
460   BL=BL-I+S                      'ADJUST BACKLOG
470   I=0
480   GOTO 590                       'PREPARE FOR TOMORROW
490   SB=BL                          'SHIP ENTIRE BACKLOG
500   I=I-BL                         'ADJUST INVENTORY
510   BL=0
520   IF I>=S THEN 570               'IF WE CAN FILL TODAY'S ORDER
530   SN=I                          'SHIP WHAT WE CAN
540   BL=S-I                         'CREATE BACKLOG
550   I=0
560   GOTO 590                       'PREPARE FOR TOMORROW
570   SN=S                          'FILL TODAY'S ORDER
580   I=I-S                          'ADJUST INVENTORY
590   IF M$="yes" THEN 650           'IF MFG HAS BEEN TRIGGERED
600   IF I<=TM THEN 630              'IF MFG IS TO BE TRIGGERED
610   M$="no": P$="no"
620   GOTO 700
```

Figure 14-2. (continued)

```
630    M$="yes": DA=1
640    GOTO 700
650    IF DA>=TR THEN 680              'IF STARTUP TIME COMPLETE
660    DA=DA+1
670    GOTO 700
680    P$="yes"
690    IF I>=SM THEN 610               'IF WE SHOULD STOP PRODUCTION
700    PRINT TAB(29) USING U$;S;
710    PRINT TAB(37) USING U$;SB;
720    PRINT TAB(47) USING U$;SN;
730    PRINT TAB(57) USING U$;BL;
740    PRINT TAB(67) USING U$;I
750 NEXT D
760 INPUT "Enter C to continue; any letter to start over ",E$
770 IF E$="C" THEN DB=DB+DS: GOTO 190 ELSE 110
```

Figure 14-3.   Simulation to show the relation between machine setting and cost.

```
10 '*************************************************************
20 '** SIMCUT.BAS   Simulation of cost when goal measurement is  **
30 '** set and there are scrap and rework costs for pieces       **
40 '** outside lower and upper cutoffs.  For Handbook of         **
50 '** Manufacturing and Production Management Formulas,         **
60 '** Charts, and Tables                                        **
70 '*************************************************************
100 CLEAR,,,32768!: SCREEN 6: COLOR 3,5
110 U1$="##": U2$="#.####": U3$="#####.##"
200 N=5000
210 FOR SELECT=1 TO 21
220    ON SELECT GOSUB 2220,2230,2240,2250,2260,2270,2280,2290,2300,2310,2320,233
0,2340,2350,2360,2370,2380,2390,2400,2410,2420
230    GOSUB 3010                'To select random numbers and do simulation
240    GOSUB 4010                'Print results for one run of 5,000
250 NEXT SELECT
2220 BR1=.1587: BR2=.9773: RETURN      '1.1240
2230 BR1=.1469: BR2=.9744: RETURN      '1.1241
2240 BR1=.1357: BR2=.9713: RETURN      '1.1242
2250 BR1=.1251: BR2=.9678: RETURN      '1.1243
2260 BR1=.1151: BR2=.9641: RETURN      '1.1244
2270 BR1=.1056: BR2=.9599: RETURN      '1.1245
2280 BR1=.0968: BR2=.9554: RETURN      '1.1246
2290 BR1=.0885: BR2=.9505: RETURN      '1.1247
2300 BR1=.0808: BR2=.9452: RETURN      '1.1248
2310 BR1=.0735: BR2=.9394: RETURN      '1.1249
2320 BR1=.0668: BR2=.9332: RETURN      '1.1250
2330 BR1=.0606: BR2=.9265: RETURN      '1.1251
2340 BR1=.0548: BR2=.9192: RETURN      '1.1252
2350 BR1=.0495: BR2=.9115: RETURN      '1.1253
2360 BR1=.0446: BR2=.9032: RETURN      '1.1254
2370 BR1=.0401: BR2=.8944: RETURN      '1.1255
2380 BR1=.0359: BR2=.8849: RETURN      '1.1256
2390 BR1=.0322: BR2=.8749: RETURN      '1.1257
2400 BR1=.0287: BR2=.8643: RETURN      '1.1258
2410 BR1=.0256: BR2=.8531: RETURN      '1.1259
2420 BR1=.0227: BR2=.8413: RETURN      '1.1260
3000 'Subroutine to select random numbers and calculate cost  $$$$$$$$$$$$$
3010 SUMCOST=0
3012 CLS
3020 FOR C=1 TO N
3022    LOCATE 12,38
3024    PRINT C
```

Figure 14-3. (continued)

```
3030    RN=INT(RND*10000)+1
3040    IF RN<BR1*10000 THEN COST=23.35 ELSE 3070
3050    SUMCOST=SUMCOST+COST
3060    GOTO 3090
3070    IF RN>BR2*10000 THEN COST=10.8 ELSE 3090
3080    SUMCOST=SUMCOST+COST
3090 NEXT C
3100 RETURN
4000 'Subroutine to print on printer    $$$$$$$$$$$$$$$$$$$
4010 LPRINT TAB(2)USING U1$;SELECT;
4020 LPRINT TAB(10)USING U2$;1.124+(SELECT-1)*.0001;
4030 LPRINT TAB(30)USING U3$;SUMCOST
4040 RETURN
```

increase and the tool must be replaced sooner. On the other hand, reducing sharpening and replacement costs raises the cost of rejected products. What is the optimum policy?

The program listed in Figure 14-4 assumes that records are available to show the reject rate as a function of the number of items produced since sharpening. Those numbers result in the cumulative distribution in lines 550 to 650.

The program starts by asking for values for variables such as the cost of a reject, the cost of sharpening, and how often will the tool be sharpened? Table 14-6 shows the results of making forty items when it costs $12.34 to reject one, $3.68 to sharpen the tool, $8.91 to replace the tool, and the tool is sharpened after making five items. The program assumes that the tool will be replaced after it is sharpened seven times.

For each item produced the table shows if it was accepted (A), if the tool was sharpened (S), if the tool was replaced (R), the added cost, and the accumulation of added costs.

MANAGEMENT SIMULATION GAMES

Simulations find another application in training managers. Experienced as well as student managers benefit from "playing" such games and studying the results of various policy actions.

One reason for management games is that a manager (or a large group of managers) can gain "experience" with a certain situation. The reason may be that simulating a situation costs much less. It may also be to study a situation that is critical but seldom arises to give actual experience.

The programs in this chapter can be used for management games. However, it is likely that a session leader would want to modify them to provide for intervention by the leader or the managers. For example, if a session is looking at the decision on how often to sharpen the cutting tool, the leader would want to change the program so that a manager could intervene at any time and change the variables.

Figure 14-5 is an computer program for a management game that challenges participants to find the production rate that results in the best long-run

Figure 14-4.   Computer program simulating cost as a function of time between tool sharpenings.

```
10 '*******************************************************
20 '** SIMTOOL  Simulates production of a product, with  **
30 '** sharpening and replacement of tool.               **
40 '** July 19, 1984                                     **
50 '*******************************************************
60 CLEAR,,,32768!: SCREEN 6: COLOR 3,2
70 DEFINT I
80 CLS
90 INPUT "Cost of reject ";REJCOST
100 INPUT "Cost of sharpening ";SHARPCOST
110 INPUT "Cost of replacing tool ";REPLCOST
120 INPUT "How often will tool be sharpened ";NBRTOSHARP
130 INPUT "Do you want a full table (Y/N) ";T$
140 IF T$="Y" THEN 160
150 INPUT "How many items will be produced ";NMAX
160 THREE=0: SINCESHARP=1: SINCEREPL=0: CUMCOST=0: ITEM=0: NEEDREPL=7
170 CLS
180 FOR C=0 TO 1
190    PRINT TAB(1+38*C)"Nbr  A S R    Cost        Cum";
200 NEXT C
210 PRINT
220 ITEM=ITEM+1                          'make a piece
230 GOSUB 550                            'determine accept or reject
240 SINCESHARP=SINCESHARP+1              'increment tool usage
250 IF SINCESHARP>NBRTOSHARP THEN 280    'if tool due for sharpening
260 S$="N": R$="N"
270 IF T$="Y" THEN 390 ELSE 520
280 IF SINCEREPL>=NEEDREPL THEN 360      'if tool due for replacement
290 'SHARPEN TOOL                        <<<<<<<<<<<<<<<<
300 R$="N": S$="Y"
310 SINCESHARP=1
320 SINCEREPL=SINCEREPL+1
330 COST=COST+SHARPCOST
340 CUMCOST=CUMCOST+SHARPCOST
350 IF T$="Y" THEN 390 ELSE 520
360 GOSUB 740                            'replace tool
370 IF T$="N" THEN 520
380 'PRINT TABLE                         <<<<<<<<<<<<<<<<<<
390 PRINT TAB(1+38*THREE) USING "####"; ITEM;
400 PRINT TAB(6+38*THREE) A$;" ";S$;" ";R$;
410 PRINT TAB(13+38*THREE) USING "###.##"; COST;
420 PRINT TAB(22+38*THREE) USING "#####.##": CUMCOST;
430 IF T$="N" THEN 530
440 IF THREE<1 THEN THREE=THREE+1: GO'       'if line not finished
450 THREE=0
460 IF ITEM/40=INT(ITEM/40) THEN 490         screen full
470 PRINT
480 GOTO 220
490 PRINT
500 INPUT "Press <Enter> for next scr
510 GOTO 170
520 IF ITEM>=NMAX THEN 390 ELSE 220
530 END
540 'Select rejection rate as functio       e since sharpening  <<<<<<
550 IF SINCESHARP<6 THEN REJ=2: GOTO
560 IF SINCESHARP<9 THEN REJ=4: GOTO
570 IF SINCESHARP<12 THEN REJ=6: GOTO
580 IF SINCESHARP<14 THEN REJ=8: GOTO
590 IF SINCESHARP<15 THEN REJ=10: GOT(
600 IF SINCESHARP<16 THEN REJ=12: GOT(
610 IF SINCESHARP<17 THEN REJ=14: GOT(
620 IF SINCESHARP<18 THEN REJ=16: GOT(
630 IF SINCESHARP<19 THEN REJ=20: GOT(
```

Figure 14-4. (continued)

```
640 IF SINCESHARP<20 THEN REJ=24: GOTO 670
650 REJ=30
660 'Determine accept or reject  <<<<<<<<<<<<<<<
670 IF INT(RND*100)+1<REJ THEN 700          'if rejected
680 A$="Y": COST=0
690 GOTO 720
700 A$="N": COST=REJCOST
710 CUMCOST=CUMCOST+REJCOST
720 RETURN
730 'Subroutine REPL replaces tool if needed        $$$$$$$$$$$$$$
740 COST=COST+REPLCOST
750 CUMCOST=CUMCOST+REPLCOST
760 SINCEREPL=0
770 'Get life of new tool                     <<<<<<<<<<<<<<
780 RANDNO=INT(RND*100)+1
790 IF RANDNO<2 THEN NEEDREPL=3: GOTO 880
800 IF RANDNO<5 THEN NEEDREPL=4: GOTO 880
810 IF RANDNO<10 THEN NEEDREPL=5: GOTO 880
820 IF RANDNO<22 THEN NEEDREPL=6: GOTO 880
830 IF RANDNO<42 THEN NEEDREPL=7: GOTO 880
840 IF RANDNO<67 THEN NEEDREPL=8: GOTO 880
850 IF RANDNO<86 THEN NEEDREPL=9: GOTO 880
860 IF RANDNO<97 THEN NEEDREPL=10: GOTO 880
870 NEEDREPL=11
880 R$="Y": S$="N"
890 SINCESHARP=1
900 RETURN
```

Figure 14-5. Management game based on simulated daily sales.

```
100 '************************************************************
110 '** PRODGAME.BAS  Management game to experiment with levels  **
120 '** of production.  For Production Management Handbook       **
130 '**                                                          **
140 '************************************************************
150 CLS
160 U3$="####":U4$="#####":U5$="######":U6$="#######"
170 DAY=0:INVENTORY=0:BACKLOG=0
180 PPU=10:CPUS=1:CPUU=15:CNGCHG=0
190 INPUT"Enter daily production rate ",PRODRATE
200 PRINT TAB(17)"--Inventory--  ---Backlog---  ---Charges---  ------Profit-----
-"
210 PRINT"Day Prod Sales  In  Out  End   In  Out  End   Inv Bklog Chng Sales   N
et   Cum"
220 DAY=DAY+1:ADDINV=0:SUBINV=0:ADDLOG=0:SUBLOG=0
230 GOSUB 300              'Find day's sales
240 IF PRODRATE>SALES THEN GOSUB 440 ELSE GOSUB 620
250 GOSUB 800             'Calculate and print
260 GOSUB 960             'Producation rate change
270 GOTO 220              'Another day
280 END
290 'Subroutine to produce day's sales  <<<<<<<<<<<<<
300 S=INT(RND*69+1)
310 IF S<2 THEN SALES=INT(RND*20+1)+819:GOTO 420
320 IF S<5 THEN SALES=INT(RND*15+1)+839:GOTO 420
330 IF S<11 THEN SALES=INT(RND*15+1)+854:GOTO 420
340 IF S<20 THEN SALES=INT(RND*10+1)+869:GOTO 420
350 IF S<30 THEN SALES=INT(RND*10+1)+879:GOTO 420
360 IF S<41 THEN SALES=INT(RND*20+1)+889:GOTO 420
370 IF S<51 THEN SALES=INT(RND*10+1)+909:GOTO 420
380 IF S<60 THEN SALES=INT(RND*10+1)+919:GOTO 420
390 IF S<66 THEN SALES=INT(RND*15+1)+929:GOTO 420
400 IF S<69 THEN SALES=INT(RND*15+1)+944:GOTO 420
```

Figure 14-5. (continued)

```
410 SALES=INT(RND*21+1)+959
420 RETURN
430 'Subroutine for days when production exceeds sales  <<<<<<<<<
440 EXCESS=PRODRATE-SALES
450 IF BACKLOG>0 THEN 500
460 SHIP=SALES                    'There is no backlog
470 ADDINV=EXCESS                 'Amount to add to inventory
480 INVENTORY=INVENTORY+ADDINV
490 GOTO 600
500 IF BACKLOG>EXCESS THEN 570    'Jp from 450 if there is a backlog
510 SHIP=SALES+BACKLOG            'Excess wipes out backlog
520 SUBLOG=BACKLOG
530 ADDINV=EXCESS-BACKLOG
540 INVENTORY=INVENTORY+ADDINV
550 BACKLOG=0
560 GOTO 600
570 SHIP=PRODRATE                 'Excess reduces backlog
580 SUBLOG=EXCESS
590 BACKLOG=BACKLOG-SUBLOG
600 RETURN
610 'Subroutine for days when there is a production shortfall  <<<<<<<<<
620 DAYSHORT=SALES-PRODRATE       'Find day's shortfall
630 IF INVENTORY>0 THEN 680       'If there is an inventory
640 SHIP=PRODRATE                 'Ship today's entire production
650 ADDLOG=DAYSHORT               'Day' shortfall will be added to backlog
660 BACKLOG=BACKLOG+ADDLOG
670 GOTO 780
680 IF INVENTORY>DAYSHORT THEN 750  'Jp from 630 if there is an inventory
690 SHIP=PRODRATE+INVENTORY       'Shortfall wipes out inventory
700 SUBINV=INVENTORY
710 ADDLOG=DAYSHORT-INVENTORY     'Amount that will be added to backlog
720 BACKLOG=BACKLOG+ADDLOG
730 INVENTORY=0
740 GOTO 780
750 SHIP=SALES                    'Today's sales are shipped
760 SUBINV=DAYSHORT               'Reduce inventory by amount of shortfall
770 INVENTORY=INVENTORY-SUBINV
780 RETURN
790 'Subroutine to calculate charges and profit, and to print  <<<<<<<
800 INVCHG=INVENTORY*CPUS
810 BKLOGCHG=BACKLOG*CPUU
820 SPROFIT=SHIP*PPU
830 NETPROFIT=SPROFIT-INVCHG-BKLOGCHG-CNGCHG
840 CUMPROFIT=CUMPROFIT+NETPROFIT
850 PRINT USING U3$;DAY;PRODRATE;
860 PRINT USING U4$;SALES;ADDINV;SUBINV;INVENTORY;ADDLOG;SUBLOG;BACKLOG;
870 PRINT USING U5$;INVCHG;BKLOGCHG;
880 PRINT USING U4$;CNGCHG;
890 PRINT" ";
900 PRINT USING U4$;SPROFIT;
910 PRINT USING U5$;NETPROFIT;
920 PRINT USING U6$;CUMPROFIT
930 CNGCHG=0
940 RETURN
950 'Subroutine to change production rate  <<<<<<<<<<<
960 LNBR=CSRLIN
970 LOCATE 25,1
980 INPUT;"Do you want to change the daily production rate? (Y/N) ",Q$
990 IF Q$="N" OR Q$="n" THEN 1040
1000 IF Q$="Y" OR Q$="y" THEN 1020
1010 PRINT"Answer Y or N only":GOTO 980
1020 LOCATE 25,1:PRINT SPACE$(56);:LOCATE 25,1:INPUT;"What should the new rate b
e? ",PRODRATE
1030 CNGCHG=500
1040 LOCATE LNBR,1
1050 RETURN
```

Table 14-6. Simulation of cost as a function of time between tool sharpenings.

```
RUN
Cost of reject? 12.34
Cost of sharpening? 3.68
Cost of replacing tool? 8.91
How often will tool be sharpened? 5
Do you want a full table (Y/N)? Y
```

Nbr	A	S	R	Cost	Cum	Nbr	A	S	R	Cost	Cum
1	Y	N	N	0.00	0.00	2	Y	N	N	0.00	0.00
3	Y	N	N	0.00	0.00	4	Y	N	N	0.00	0.00
5	Y	Y	N	3.68	3.68	6	Y	N	N	0.00	3.68
7	Y	N	N	0.00	3.68	8	Y	N	N	0.00	3.68
9	Y	N	N	0.00	3.68	10	Y	Y	N	3.68	7.36
11	Y	N	N	0.00	7.36	12	Y	N	N	0.00	7.36
13	Y	N	N	0.00	7.36	14	Y	N	N	0.00	7.36
15	Y	Y	N	3.68	11.04	16	Y	N	N	0.00	11.04
17	N	N	N	12.34	23.38	18	Y	N	N	0.00	23.38
19	Y	N	N	0.00	23.38	20	Y	Y	N	3.68	27.06
21	Y	N	N	0.00	27.06	22	Y	N	N	0.00	27.06
23	Y	N	N	0.00	27.06	24	Y	N	N	0.00	27.06
25	Y	Y	N	3.68	30.74	26	Y	N	N	0.00	30.74
27	Y	N	N	0.00	30.74	28	Y	N	N	0.00	30.74
29	N	N	N	12.34	43.08	30	Y	Y	N	3.68	46.76
31	Y	N	N	0.00	46.76	32	Y	N	N	0.00	46.76
33	Y	N	N	0.00	46.76	34	Y	N	N	0.00	46.76
35	Y	Y	N	3.68	50.44	36	Y	N	N	0.00	50.44
37	Y	N	N	0.00	50.44	38	Y	N	N	0.00	50.44
39	Y	N	N	0.00	50.44	40	Y	N	Y	8.91	59.35

Press < Enter> for next screen

Nbr	A	S	R	Cost	Cum	Nbr	A	S	R	Cost	Cum
41	Y	N	N	0.00	59.35	42	Y	N	N	0.00	59.35
43	Y	N	N	0.00	59.35	44	Y	N	N	0.00	59.35
45	Y	Y	N	3.68	63.03	46	Y	N	N	0.00	63.03
47	Y	N	N	0.00	63.03	48	N	N	N	12.34	75.37
49	Y	N	N	0.00	75.37	50	Y	Y	N	3.68	79.05
51	Y	N	N	0.00	79.05	52	Y	N	N	0.00	79.05
53	Y	N	N	0.00	79.05	54	Y	N	N	0.00	79.05
55	Y	Y	N	3.68	82.73	56	Y	N	N	0.00	82.73
57	Y	N	N	0.00	82.73	58	Y	N	N	0.00	82.73
59	Y	N	N	0.00	82.73	60	Y	Y	N	3.68	86.41
61	Y	N	N	0.00	86.41	62	Y	N	N	0.00	86.41
63	Y	N	N	0.00	86.41	64	Y	N	N	0.00	86.41
65	Y	Y	N	3.68	90.09	66	Y	N	N	0.00	90.09
67	Y	N	N	0.00	90.09	68	Y	N	N	0.00	90.09
69	Y	N	N	0.00	90.09	70	Y	Y	N	3.68	93.77
71	Y	N	N	0.00	93.77	72	Y	N	N	0.00	93.77
73	Y	N	N	0.00	93.77	74	Y	N	N	0.00	93.77
75	Y	Y	N	3.68	97.45	76	Y	N	N	0.00	97.45
77	Y	N	N	0.00	97.45	78	Y	N	N	0.00	97.45
79	Y	N	N	0.00	97.45	80	Y	N	Y	8.91	106.36

Press < Enter> for next screen
Break in 50
0

cumulative profit. For each day simulated, the computer prints the number produced, the number sold, the numbers going into and out of inventory and backlog, and the financial results. At the end of each day the computer stops so the participant can examine the numbers and make a decision about production rates.

Participants are given the following facts and rules:

1. Your company is producing a new version of a product that was averaging 800 sales per day.
2. You make a profit of $10 on each unit when it is sold.
3. It costs $1 per day to keep a unit in inventory.
4. It costs $15 per day to keep a unit on a backlog list.
5. You may change the production rate (number manufactured per day) at any time, but it costs $500 to make the change.

The backlog charge represents losses such as customers buying from a competitor and not returning next time they need the item. Sophistication can be added to the program by replacing the fixed $15 charge with a function that depresses sales and allows them to recover gradually.

Management classes can use this game to give practice in interpreting daily results. Competitors can be individuals or teams, and their cumulative profits can be compared after a period such as sixty days.

Appendix

TABLE A-1
AREA UNDER NORMAL CURVE

Statisticians named the outcome that is the result of serveral random influences a normal distribution. Its graph is a normal curve. For example, the speed with which assembly line workers splice wires forms a normal curve because it shows individual abilities that in turn are affected by heredity, training, room temperature, health, recent meals, and many other factors.

Most practical applications look at the area under a portion of the curve, while considering that the area under the entire curve is one. There are several ways that this area can be presented; Table A-1 gives the area to the right of the midpoint, measured in standard deviations or Z values. The standard deviation's first decimal place is given in the Z column, and the second place is given at the top of each column. To find the area between the midpoint and 0.25 standard deviations, look down the Z column to 0.2 and then go across that row to the column headed .05. At that intersection, note that the area between the midpoint and 0.25 standard deviations to its right is 0.0987.

Because the normal curve is symmetrical, you can use Table A-1 to find areas between any two points, area below a point, or area above a point. For example, half the area lies below the midpoint and half above it. Therefore, to find the area below 0.25 standard deviations above the midpoint, add 0.5 to the number you just found — the answer is 0.5987.

Instructions for using this and other tables in the appendix are included where needed throughout the book.

Table A-1. Area under normal curve.
Area Under Standard Normal Curve

Z	.00	.01	.02	.03	.04	.05	.06	.07	.08	.09
0.0	.0000	.0040	.0080	.0120	.0160	.0199	.0239	.0279	.0319	.0359
0.1	.0398	.0438	.0478	.0517	.0557	.0596	.0636	.0675	.0714	.0753
0.2	.0793	.0832	.0871	.0910	.0948	.0987	.1026	.1064	.1103	.1141
0.3	.1179	.1217	.1255	.1293	.1331	.1368	.1406	.1443	.1480	.1517
0.4	.1554	.1591	.1628	.1664	.1700	.1736	.1772	.1808	.1844	.1879
0.5	.1915	.1950	.1985	.2019	.2054	.2088	.2123	.2157	.2190	.2224
0.6	.2257	.2291	.2324	.2357	.2389	.2422	.2454	.2486	.2517	.2549
0.7	.2580	.2611	.2642	.2673	.2704	.2734	.2764	.2794	.2823	.2852
0.8	.2881	.2910	.2939	.2967	.2995	.3023	.3051	.3078	.3106	.3133
0.9	.3159	.3186	.3212	.3238	.3264	.3289	.3315	.3340	.3365	.3389
1.0	.3413	.3438	.3461	.3485	.3508	.3531	.3554	.3577	.3599	.3621
1.1	.3643	.3665	.3686	.3708	.3729	.3749	.3770	.3790	.3810	.3830
1.2	.3849	.3869	.3888	.3907	.3925	.3944	.3962	.3980	.3997	.4015
1.3	.4032	.4049	.4066	.4082	.4099	.4115	.4131	.4147	.4162	.4177
1.4	.4192	.4207	.4222	.4236	.4251	.4265	.4279	.4292	.4306	.4319
1.5	.4332	.4345	.4357	.4370	.4382	.4394	.4406	.4418	.4429	.4441
1.6	.4452	.4463	.4474	.4484	.4495	.4505	.4515	.4525	.4535	.4545
1.7	.4554	.4564	.4573	.4582	.4591	.4599	.4608	.4616	.4625	.4633
1.8	.4641	.4649	.4656	.4664	.4671	.4678	.4686	.4693	.4699	.4706
1.9	.4713	.4719	.4726	.4732	.4738	.4744	.4750	.4756	.4761	.4767
2.0	.4772	.4778	.4783	.4788	.4793	.4798	.4803	.4808	.4812	.4817
2.1	.4821	.4826	.4830	.4834	.4838	.4842	.4846	.4850	.4854	.4857
2.2	.4861	.4864	.4868	.4871	.4875	.4878	.4881	.4884	.4887	.4890
2.3	.4893	.4896	.4898	.4901	.4904	.4906	.4909	.4911	.4913	.4916
2.4	.4918	.4920	.4922	.4925	.4927	.4929	.4931	.4932	.4934	.4936
2.5	.4938	.4940	.4941	.4943	.4945	.4946	.4948	.4949	.4951	.4952
2.6	.4953	.4955	.4956	.4957	.4959	.4960	.4961	.4962	.4963	.4964
2.7	.4965	.4966	.4967	.4968	.4969	.4970	.4971	.4972	.4973	.4974
2.8	.4974	.4975	.4976	.4977	.4977	.4978	.4979	.4979	.4980	.4981
2.9	.4981	.4982	.4982	.4983	.4984	.4984	.4985	.4985	.4986	.4986
3.0	.4987	.4987	.4987	.4988	.4988	.4989	.4989	.4989	.4990	.4990

TABLE A-2
4800 RANDOM DIGITS

Humans do not pick numbers with true randomness; we fall into patterns and show biases. Therefore it is best to use a random number table such as Table A-2 for simulation and other applications. Most computers have routines that generate series of random numbers.

Table A-2 includes an analysis of its randomness. To be random, every digit 0 to 9 must have an equal chance of being picked at any time. For example, a 2 is just as likely to be picked after a 2 as it is after a 9 or a 3. The end of Table A-2 shows how many times each digit appears in the table, and how many times each digit follows any given digit.

Table A-2. 4800 random digits and analysis of randomness.

Line											
1	76496	13519	23285	40803	55138	84706	47058	81511	33915	86108	03605	06477
2	96257	65891	20708	31000	20350	82072	39102	56042	00514	90705	62758	97812
3	02146	64876	57220	53153	12565	36661	37552	23494	05968	68155	22426	25266
4	76743	15235	62765	66299	02130	49275	88244	44915	58626	04061	20723	78863
5	20328	43770	90048	57665	93746	38395	47071	97751	33077	23015	04475	08852
6	35540	71992	99183	87318	76642	36824	53979	32041	30682	69097	09695	72299
7	94777	25575	95905	15775	43327	67354	43961	48949	52729	37600	83118	71401
8	08290	08329	56527	10813	96231	25575	72843	75784	81857	50346	83947	63555
9	69658	36826	08654	72958	16752	46181	78757	11512	18547	23107	35099	08768
10	59241	41979	98322	31247	07342	43763	08409	27580	45967	91464	87295	66297
11	71617	49269	63542	84838	59487	41122	99014	61025	75898	68140	84663	25288
12	15046	55498	70142	73067	65539	64952	87305	14084	89487	58816	66014	55878
13	74197	81148	46024	96971	15977	16942	30296	43813	89215	86275	97475	06215
14	50240	64131	28323	92717	14259	13687	84893	10640	46751	76927	85759	67735
15	34854	17029	72843	52522	41856	88976	66930	93631	84275	43756	07971	87595
16	08299	54543	16510	44742	19146	55413	33620	51914	73325	10352	99954	46632
17	75043	80456	16356	18425	00590	30789	32689	74735	25391	92338	89605	13382
18	35557	40379	99177	62717	06565	36686	90040	01341	23753	14566	24066	67813
19	62300	47874	02097	38846	28633	88816	73708	32879	68950	80457	60032	58332
20	86760	07874	01741	13068	49161	00951	19798	06475	42471	35984	44648	55026
21	51198	43849	15432	57419	60420	07612	55834	12297	57786	62568	03607	37090
22	96163	89592	80885	31248	74189	13771	46040	10572	85044	16181	94135	75513
23	96154	31705	72904	22691	71628	42721	11330	79556	38044	81224	53056	26711
24	30420	22180	18835	80804	55384	18584	26620	69477	97925	02591	29184	62930
25	12651	68218	46270	95035	16669	47610	18843	59380	55478	94883	56543	60239
26	33186	71891	12246	17400	38111	42832	14893	18334	23752	60781	94058	81743
27	95215	88712	72775	14076	10010	58722	80715	15955	83335	35087	77723	03719
28	60507	91472	55672	65381	56916	70182	64304	21099	28788	34284	44483	87485
29	50412	36763	07962	42270	48206	30604	92604	03904	48479	70183	41152	91908
30	66318	27286	74569	91092	77342	45089	22419	33727	46668	78566	25834	05513
31	71785	88715	04398	14092	35702	28809	99954	79794	15035	42825	22364	88359
32	86983	23859	23372	59298	33855	94877	32852	90152	34858	28052	80646	00561
33	83581	68293	08417	96957	55051	77773	17219	66069	17778	18569	56604	30308
34	74958	35747	67646	94325	82880	81201	60978	57640	46983	38228	86373	37905
35	31453	31883	66775	22539	35701	41061	29945	31263	21043	12849	08748	71419
36	77677	08303	31826	50836	86092	61636	41243	68846	36327	65817	67552	31178
37	81958	52919	21725	58264	48115	76928	62697	27743	72385	00021	02929	77238
38	58796	97293	81576	84178	82737	90519	39101	88104	30515	52164	78766	67836
39	88530	57036	58914	38616	10018	27209	80709	01264	14259	36849	14894	72110
40	53669	46841	70143	50805	24548	02582	71307	44160	03003	65746	75330	60325
41	76643	90600	84986	18101	17676	39259	48828	96059	02300	55568	88098	22609
42	27932	02824	35260	93562	68881	52902	39286	18288	91446	39097	01911	05299
43	00924	26275	72007	53214	72634	66423	81407	79785	82873	12824	60894	71341
44	15069	48566	30202	68504	48392	85413	41314	37915	67188	29662	13952	18093
45	46737	81387	77801	87589	61446	54465	54902	73905	18523	63181	95929	75691
46	72496	89431	20716	00487	20334	67371	60125	56803	47675	69005	55810	42382
47	18617	49490	25843	76453	29557	58722	13878	55956	45805	52995	38647	88103
48	53668	92166	30135	74744	48454	32320	41174	39390	95471	35653	81347	53401
49	81028	45140	91588	64588	00513	23867	92759	59382	19154	24890	33614	46223
50	03258	11124	36015	15932	67105	36825	21817	02049	78122	52089	48671	97906

Line

51	01617 24891 10552 06222 41789 04116 75191 39648 75035 45131 46364 94596
52	77684 90842 61911 98505 33924 32412 73698 20306 10172 95739 81576 17230
53	12728 52088 46118 49128 16999 45235 79459 42290 86093 48574 11152 06476
54	19319 05993 82178 48018 90274 67455 22262 56726 85977 09158 63296 50050
55	70916 63267 35914 45522 59319 02677 69278 32861 99573 80463 84733 17419
56	86808 60603 23440 20906 94692 85490 35074 38853 10172 28891 11577 79700
57	29736 12902 22502 31288 17683 30798 78903 44838 49231 26344 10542 84160
58	11788 42747 59184 79624 99641 52071 45670 19033 78120 08303 19774 10735
59	24523 55739 71229 82452 43257 64123 82008 22624 90410 99771 53055 49973
60	60421 84550 25833 50828 40778 01644 27314 44930 95489 04039 81321 39700
61	21942 35902 38749 33889 83675 78226 17210 43906 87786 56190 49606 69384
62	89564 42686 66952 09861 83337 99763 08731 89970 47682 61534 94995 89050
63	65548 09276 57307 38913 02303 89654 36792 31018 76575 52064 53031 89957
64	31352 61472 47988 99280 63162 71883 41306 69538 67094 33353 82939 17331
65	99566 12087 84749 32110 46975 60841 86389 41506 01530 40021 19613 53239
66	42549 96592 04584 46649 43410 90440 91655 58358 91351 76850 01989 67997
67	70761 17899 64229 91915 57857 67362 11338 48923 38027 67533 83979 27572
68	77591 91480 01997 35373 70855 93198 94142 91777 04018 35662 04400 93492
69	53598 51158 52502 49972 83683 14541 87202 67836 01692 87037 10474 45614
70	81932 02593 73869 92948 99712 44202 15433 25357 31439 45242 48826 42373
71	72392 79408 02913 52547 98619 98054 37647 53478 75898 45088 45661 63728
72	08138 84474 85758 80997 64855 79498 89851 22546 26240 71047 67723 78185
73	82728 44394 69118 65265 17590 22227 17988 81506 86162 87020 42007 45530
74	05633 72679 82008 55887 20411 51131 60963 19897 46815 77610 16527 35381
75	49232 93182 80551 22690 04870 72723 73790 86554 98967 67518 46126 27594
76	16983 21534 00373 43051 32154 17875 04313 51845 34879 89517 50790 60317
77	07267 90526 08587 88288 16814 92098 79527 03997 57830 40108 04474 44077
78	05632 95722 12009 17256 37429 12165 45357 02957 47599 52173 24080 37829
79	02479 70952 89852 99383 96258 19677 50716 17161 07335 51134 69424 70801
80	08046 30789 55741 04836 87761 19336 40629 08776 28627 41053 74630 61161

	0	1	2	3	4	5	6	7	8	9
Frequency	468	485	481	463	465	500	449	504	520	465

					Following					
	0	1	2	3	4	5	6	7	8	9
Digit										
0	44	48	47	51	40	46	47	45	44	56
1	45	45	40	53	55	41	55	41	55	55
2	48	46	42	47	46	57	43	58	47	47
3	41	50	55	43	48	48	41	47	51	39
4	62	48	44	33	42	45	41	48	59	43
5	50	50	49	57	39	59	38	61	51	46
6	38	42	49	39	47	46	47	57	42	42
7	42	49	50	44	51	57	51	52	57	50
8	54	57	54	55	53	55	44	50	58	40
9	44	49	51	41	44	46	42	45	56	47

TABLE A-3
BINOMIAL PROBABILITY

Binomial, or two-state, probability applies when exactly two outcomes are possible. It applies when you consider an assembly line as either operating or shut down. It would not apply if you considered that the line could be operating at capacity, at seventy-five percent, at fifty percent, or shut down.

One example will demonstrate how to use the table. Suppose your project buys quartz controlled clocks for control panels, and incoming inspection notes that ten percent of them arrive set to the correct time. What is the probability that, out of the next five, three of them will be set to the correct time?

For any one clock, the probability of being set is ten percent (.10); locate that number in the column headings. Then look down the n column to 5. In the adjacent group of numbers in the x column, locate 3. Go across that row to the column headed .10, and read .0081 at that intersection. That is the probability of exactly three clocks out of the next five being set to the correct time.

More explanation and examples are given in appropriate sections of the book.

Table A-3. Table of binomial probability.

						p					
n	x	.01	.02	.05	.10	.15	.20	.25	.30	.40	.50
1	0	.9900	.9800	.9500	.9000	.8500	.8000	.7500	.7000	.6000	.5000
	1	.0100	.0200	.0500	.1000	.1500	.2000	.2500	.3000	.4000	.5000
2	0	.9801	.9604	.9025	.8100	.7225	.6400	.5625	.4900	.3600	.2500
	1	.0198	.0392	.0950	.1800	.2550	.3200	.3750	.4200	.4800	.5000
	2	.0001	.0004	.0025	.0100	.0225	.0400	.0625	.0900	.1600	.2500
3	0	.9703	.9412	.8574	.7290	.6141	.5120	.4219	.3430	.2160	.1250
	1	.0294	.0576	.1354	.2430	.3251	.3840	.4219	.4410	.4320	.3750
	2	.0003	.0012	.0071	.0270	.0574	.0960	.1406	.1890	.2880	.3750
	3	.0000	.0000	.0001	.0010	.0034	.0080	.0156	.0270	.0640	.1250
4	0	.9606	.9224	.8145	.6561	.5220	.4096	.3164	.2401	.1296	.0625
	1	.0388	.0753	.1715	.2916	.3685	.4096	.4219	.4116	.3456	.2500
	2	.0006	.0023	.0135	.0486	.0975	.1536	.2109	.2646	.3456	.3750
	3	.0000	.0000	.0005	.0036	.0115	.0256	.0469	.0756	.1536	.2500
	4	.0000	.0000	.0000	.0001	.0005	.0016	.0039	.0081	.0256	.0625
5	0	.9510	.9039	.7738	.5905	.4437	.3277	.2373	.1681	.0778	.0313
	1	.0480	.0922	.2036	.3280	.3915	.4096	.3955	.3601	.2592	.1563
	2	.0010	.0038	.0214	.0729	.1382	.2048	.2637	.3087	.3456	.3125
	3	.0000	.0001	.0011	.0081	.0244	.0512	.0879	.1323	.2304	.3125
	4	.0000	.0000	.0000	.0004	.0022	.0064	.0146	.0284	.0768	.1563
	5	.0000	.0000	.0000	.0000	.0001	.0003	.0010	.0024	.0102	.0313
6	0	.9415	.8858	.7351	.5314	.3771	.2621	.1780	.1176	.0467	.0156
	1	.0571	.1085	.2321	.3543	.3993	.3932	.3560	.3025	.1866	.0938
	2	.0014	.0055	.0305	.0984	.1762	.2458	.2966	.3241	.3110	.2344
	3	.0000	.0002	.0021	.0146	.0415	.0819	.1318	.1852	.2765	.3125
	4	.0000	.0000	.0001	.0012	.0055	.0154	.0330	.0595	.1382	.2344
	5	.0000	.0000	.0000	.0001	.0004	.0015	.0044	.0102	.0369	.0938
	6	.0000	.0000	.0000	.0000	.0000	.0001	.0002	.0007	.0041	.0156
7	0	.9321	.8681	.6983	.4783	.3206	.2097	.1335	.0824	.0280	.0078
	1	.0659	.1240	.2573	.3720	.3960	.3670	.3115	.2471	.1306	.0547
	2	.0020	.0076	.0406	.1240	.2097	.2753	.3115	.3177	.2613	.1641
	3	.0000	.0003	.0036	.0230	.0617	.1147	.1730	.2269	.2903	.2734
	4	.0000	.0000	.0002	.0026	.0109	.0287	.0577	.0972	.1935	.2734
	5	.0000	.0000	.0000	.0002	.0012	.0043	.0115	.0250	.0774	.1641
	6	.0000	.0000	.0000	.0000	.0001	.0004	.0013	.0036	.0172	.0547
	7	.0000	.0000	.0000	.0000	.0000	.0000	.0001	.0002	.0016	.0078
8	0	.9227	.8508	.6634	.4305	.2725	.1678	.1001	.0576	.0168	.0039
	1	.0746	.1389	.2793	.3826	.3847	.3355	.2670	.1977	.0896	.0313
	2	.0026	.0099	.0515	.1488	.2376	.2936	.3115	.2965	.2090	.1094
	3	.0001	.0004	.0054	.0331	.0839	.1468	.2076	.2541	.2787	.2188
	4	.0000	.0000	.0004	.0046	.0185	.0459	.0865	.1361	.2322	.2734
	5	.0000	.0000	.0000	.0004	.0026	.0092	.0231	.0467	.1239	.2188
	6	.0000	.0000	.0000	.0000	.0002	.0011	.0038	.0100	.0413	.1094
	7	.0000	.0000	.0000	.0000	.0000	.0001	.0004	.0012	.0079	.0313
	8	.0000	.0000	.0000	.0000	.0000	.0000	.0000	.0001	.0007	.0039

Table A-3. Table of binomial probability. (continued)

n	x	.01	.02	.05	.10	.15	.20	.25	.30	.40	.50
							p				
9	0	.9135	.8337	.6302	.3874	.2316	.1342	.0751	.0404	.0101	.0020
	1	.0830	.1531	.2985	.3874	.3679	.3020	.2253	.1556	.0605	.0176
	2	.0034	.0125	.0629	.1722	.2597	.3020	.3003	.2668	.1612	.0703
	3	.0001	.0006	.0077	.0446	.1069	.1762	.2336	.2668	.2508	.1641
	4	.0000	.0000	.0006	.0074	.0283	.0661	.1168	.1715	.2508	.2461
	5	.0000	.0000	.0000	.0008	.0050	.0165	.0389	.0735	.1672	.2461
	6	.0000	.0000	.0000	.0001	.0006	.0028	.0087	.0210	.0743	.1641
	7	.0000	.0000	.0000	.0000	.0000	.0003	.0012	.0039	.0212	.0703
	8	.0000	.0000	.0000	.0000	.0000	.0000	.0001	.0004	.0035	.0176
	9	.0000	.0000	.0000	.0000	.0000	.0000	.0000	.0000	.0003	.0020
10	0	.9044	.8171	.5987	.3487	.1969	.1074	.0563	.0282	.0060	.0010
	1	.0914	.1667	.3151	.3874	.3474	.2684	.1877	.1211	.0403	.0098
	2	.0042	.0153	.0746	.1937	.2759	.3020	.2816	.2335	.1209	.0439
	3	.0001	.0008	.0105	.0574	.1298	.2013	.2503	.2668	.2150	.1172
	4	.0000	.0000	.0010	.0112	.0401	.0881	.1460	.2001	.2508	.2051
	5	.0000	.0000	.0001	.0015	.0085	.0264	.0584	.1029	.2007	.2461
	6	.0000	.0000	.0000	.0001	.0012	.0055	.0162	.0368	.1115	.2051
	7	.0000	.0000	.0000	.0000	.0001	.0008	.0031	.0090	.0425	.1172
	8	.0000	.0000	.0000	.0000	.0000	.0001	.0004	.0014	.0106	.0439
	9	.0000	.0000	.0000	.0000	.0000	.0000	.0000	.0001	.0016	.0098
	10	.0000	.0000	.0000	.0000	.0000	.0000	.0000	.0000	.0001	.0010
11	0	.8953	.8007	.5688	.3138	.1673	.0859	.0422	.0198	.0036	.0005
	1	.0995	.1798	.3293	.3835	.3248	.2362	.1549	.0932	.0266	.0054
	2	.0050	.0183	.0867	.2131	.2866	.2953	.2581	.1998	.0887	.0269
	3	.0002	.0011	.0137	.0710	.1517	.2215	.2581	.2568	.1774	.0806
	4	.0000	.0000	.0014	.0158	.0536	.1107	.1721	.2201	.2365	.1611
	5	.0000	.0000	.0001	.0025	.0132	.0388	.0803	.1321	.2207	.2256
	6	.0000	.0000	.0000	.0003	.0023	.0097	.0268	.0566	.1471	.2256
	7	.0000	.0000	.0000	.0000	.0003	.0017	.0064	.0173	.0701	.1611
	8	.0000	.0000	.0000	.0000	.0000	.0002	.0011	.0037	.0234	.0806
	9	.0000	.0000	.0000	.0000	.0000	.0000	.0001	.0005	.0052	.0269
	10	.0000	.0000	.0000	.0000	.0000	.0000	.0000	.0000	.0007	.0054
	11	.0000	.0000	.0000	.0000	.0000	.0000	.0000	.0000	.0000	.0005
12	0	.8864	.7847	.5404	.2824	.1422	.0687	.0317	.0138	.0022	.0002
	1	.1074	.1922	.3413	.3766	.3012	.2062	.1267	.0712	.0174	.0029
	2	.0060	.0216	.0988	.2301	.2924	.2835	.2323	.1678	.0639	.0161
	3	.0002	.0015	.0173	.0852	.1720	.2362	.2581	.2397	.1419	.0537
	4	.0000	.0001	.0021	.0213	.0683	.1329	.1936	.2311	.2128	.1208
	5	.0000	.0000	.0002	.0038	.0193	.0532	.1032	.1585	.2270	.1934
	6	.0000	.0000	.0000	.0005	.0040	.0155	.0401	.0792	.1766	.2256
	7	.0000	.0000	.0000	.0000	.0006	.0033	.0115	.0291	.1009	.1934
	8	.0000	.0000	.0000	.0000	.0001	.0005	.0024	.0078	.0420	.1208
	9	.0000	.0000	.0000	.0000	.0000	.0001	.0004	.0015	.0125	.0537
	10	.0000	.0000	.0000	.0000	.0000	.0000	.0000	.0002	.0025	.0161
	11	.0000	.0000	.0000	.0000	.0000	.0000	.0000	.0000	.0003	.0029
	12	.0000	.0000	.0000	.0000	.0000	.0000	.0000	.0000	.0000	.0002

Table A-3. Table of binomial probability. (continued)

n	x	.01	.02	.05	.10	p .15	.20	.25	.30	.40	.50
13	0	.8775	.7690	.5133	.2542	.1209	.0550	.0238	.0097	.0013	.0001
	1	.1152	.2040	.3512	.3672	.2774	.1787	.1029	.0540	.0113	.0016
	2	.0070	.0250	.1109	.2448	.2937	.2680	.2059	.1388	.0453	.0095
	3	.0003	.0019	.0214	.0997	.1900	.2457	.2517	.2181	.1107	.0349
	4	.0000	.0001	.0028	.0277	.0838	.1535	.2097	.2337	.1845	.0873
	5	.0000	.0000	.0003	.0055	.0266	.0691	.1258	.1803	.2214	.1571
	6	.0000	.0000	.0000	.0008	.0063	.0230	.0559	.1030	.1968	.2095
	7	.0000	.0000	.0000	.0001	.0011	.0058	.0186	.0442	.1312	.2095
	8	.0000	.0000	.0000	.0000	.0001	.0011	.0047	.0142	.0656	.1571
	9	.0000	.0000	.0000	.0000	.0000	.0001	.0009	.0034	.0243	.0873
	10	.0000	.0000	.0000	.0000	.0000	.0000	.0001	.0006	.0065	.0349
	11	.0000	.0000	.0000	.0000	.0000	.0000	.0000	.0001	.0012	.0095
	12	.0000	.0000	.0000	.0000	.0000	.0000	.0000	.0000	.0001	.0016
	13	.0000	.0000	.0000	.0000	.0000	.0000	.0000	.0000	.0000	.0001
14	0	.8687	.7536	.4877	.2288	.1028	.0440	.0178	.0068	.0008	.0001
	1	.1229	.2153	.3593	.3559	.2539	.1539	.0832	.0407	.0073	.0009
	2	.0081	.0286	.1229	.2570	.2912	.2501	.1802	.1134	.0317	.0056
	3	.0003	.0023	.0259	.1142	.2056	.2501	.2402	.1943	.0845	.0222
	4	.0000	.0001	.0037	.0349	.0998	.1720	.2202	.2290	.1549	.0611
	5	.0000	.0000	.0004	.0078	.0352	.0860	.1468	.1963	.2066	.1222
	6	.0000	.0000	.0000	.0013	.0093	.0322	.0734	.1262	.2066	.1833
	7	.0000	.0000	.0000	.0002	.0019	.0092	.0280	.0618	.1574	.2095
	8	.0000	.0000	.0000	.0000	.0003	.0020	.0082	.0232	.0918	.1833
	9	.0000	.0000	.0000	.0000	.0000	.0003	.0018	.0066	.0408	.1222
	10	.0000	.0000	.0000	.0000	.0000	.0000	.0003	.0014	.0136	.0611
	11	.0000	.0000	.0000	.0000	.0000	.0000	.0000	.0002	.0033	.0222
	12	.0000	.0000	.0000	.0000	.0000	.0000	.0000	.0000	.0005	.0056
	13	.0000	.0000	.0000	.0000	.0000	.0000	.0000	.0000	.0001	.0009
	14	.0000	.0000	.0000	.0000	.0000	.0000	.0000	.0000	.0000	.0001
15	0	.8601	.7386	.4633	.2059	.0874	.0352	.0134	.0047	.0005	.0000
	1	.1303	.2261	.3658	.3432	.2312	.1319	.0668	.0305	.0047	.0005
	2	.0092	.0323	.1348	.2669	.2856	.2309	.1559	.0916	.0219	.0032
	3	.0004	.0029	.0307	.1285	.2184	.2501	.2252	.1700	.0634	.0139
	4	.0000	.0002	.0049	.0428	.1156	.1876	.2252	.2186	.1268	.0417
	5	.0000	.0000	.0006	.0105	.0449	.1032	.1651	.2061	.1859	.0916
	6	.0000	.0000	.0000	.0019	.0132	.0430	.0917	.1472	.2066	.1527
	7	.0000	.0000	.0000	.0003	.0030	.0138	.0393	.0811	.1771	.1964
	8	.0000	.0000	.0000	.0000	.0005	.0035	.0131	.0348	.1181	.1964
	9	.0000	.0000	.0000	.0000	.0001	.0007	.0034	.0116	.0612	.1527
	10	.0000	.0000	.0000	.0000	.0000	.0001	.0007	.0030	.0245	.0916
	11	.0000	.0000	.0000	.0000	.0000	.0000	.0001	.0006	.0074	.0417
	12	.0000	.0000	.0000	.0000	.0000	.0000	.0000	.0001	.0016	.0139
	13	.0000	.0000	.0000	.0000	.0000	.0000	.0000	.0000	.0003	.0032
	14	.0000	.0000	.0000	.0000	.0000	.0000	.0000	.0000	.0000	.0005
	15	.0000	.0000	.0000	.0000	.0000	.0000	.0000	.0000	.0000	.0000

Table A-3. Table of binomial probability. (continued)

n	x	.01	.02	.05	.10	.15	.20	.25	.30	.40	.50
16	0	.8515	.7238	.4401	.1853	.0743	.0281	.0100	.0033	.0003	.0000
	1	.1376	.2363	.3706	.3294	.2097	.1126	.0535	.0228	.0030	.0002
	2	.0104	.0362	.1463	.2745	.2775	.2111	.1336	.0732	.0150	.0018
	3	.0005	.0034	.0359	.1423	.2285	.2463	.2079	.1465	.0468	.0085
	4	.0000	.0002	.0061	.0514	.1311	.2001	.2252	.2040	.1014	.0278
	5	.0000	.0000	.0008	.0137	.0555	.1201	.1802	.2099	.1623	.0667
	6	.0000	.0000	.0001	.0028	.0180	.0550	.1101	.1649	.1983	.1222
	7	.0000	.0000	.0000	.0004	.0045	.0197	.0524	.1010	.1889	.1746
	8	.0000	.0000	.0000	.0001	.0009	.0055	.0197	.0487	.1417	.1964
	9	.0000	.0000	.0000	.0000	.0001	.0012	.0058	.0185	.0840	.1746
	10	.0000	.0000	.0000	.0000	.0000	.0002	.0014	.0056	.0392	.1222
	11	.0000	.0000	.0000	.0000	.0000	.0000	.0002	.0013	.0142	.0667
	12	.0000	.0000	.0000	.0000	.0000	.0000	.0000	.0002	.0040	.0278
	13	.0000	.0000	.0000	.0000	.0000	.0000	.0000	.0000	.0008	.0085
	14	.0000	.0000	.0000	.0000	.0000	.0000	.0000	.0000	.0001	.0018
	15	.0000	.0000	.0000	.0000	.0000	.0000	.0000	.0000	.0000	.0002
	16	.0000	.0000	.0000	.0000	.0000	.0000	.0000	.0000	.0000	.0000
17	0	.8429	.7093	.4181	.1668	.0631	.0225	.0075	.0023	.0002	.0000
	1	.1447	.2461	.3741	.3150	.1893	.0957	.0426	.0169	.0019	.0001
	2	.0117	.0402	.1575	.2800	.2673	.1914	.1136	.0581	.0102	.0010
	3	.0006	.0041	.0415	.1556	.2359	.2393	.1893	.1245	.0341	.0052
	4	.0000	.0003	.0076	.0605	.1457	.2093	.2209	.1868	.0796	.0182
	5	.0000	.0000	.0010	.0175	.0668	.1361	.1914	.2081	.1379	.0472
	6	.0000	.0000	.0001	.0039	.0236	.0680	.1276	.1784	.1839	.0944
	7	.0000	.0000	.0000	.0007	.0065	.0267	.0668	.1201	.1927	.1484
	8	.0000	.0000	.0000	.0001	.0014	.0084	.0279	.0644	.1606	.1855
	9	.0000	.0000	.0000	.0000	.0003	.0021	.0093	.0276	.1070	.1855
	10	.0000	.0000	.0000	.0000	.0000	.0004	.0025	.0095	.0571	.1484
	11	.0000	.0000	.0000	.0000	.0000	.0001	.0005	.0026	.0242	.0944
	12	.0000	.0000	.0000	.0000	.0000	.0000	.0001	.0006	.0081	.0472
	13	.0000	.0000	.0000	.0000	.0000	.0000	.0000	.0001	.0021	.0182
	14	.0000	.0000	.0000	.0000	.0000	.0000	.0000	.0000	.0004	.0052
	15	.0000	.0000	.0000	.0000	.0000	.0000	.0000	.0000	.0001	.0010
	16	.0000	.0000	.0000	.0000	.0000	.0000	.0000	.0000	.0000	.0001
	17	.0000	.0000	.0000	.0000	.0000	.0000	.0000	.0000	.0000	.0000

Table A-3. Table of binomial probability. (continued)

n	x	.01	.02	.05	.10	p .15	.20	.25	.30	.40	.50
18	0	.8345	.6951	.3972	.1501	.0536	.0180	.0056	.0016	.0001	.0000
	1	.1517	.2554	.3763	.3002	.1704	.0811	.0338	.0126	.0012	.0001
	2	.0130	.0443	.1683	.2835	.2556	.1723	.0958	.0458	.0069	.0006
	3	.0007	.0048	.0473	.1680	.2406	.2297	.1704	.1046	.0246	.0031
	4	.0000	.0004	.0093	.0700	.1592	.2153	.2130	.1681	.0614	.0117
	5	.0000	.0000	.0014	.0218	.0787	.1507	.1988	.2017	.1146	.0327
	6	.0000	.0000	.0002	.0052	.0301	.0816	.1436	.1873	.1655	.0708
	7	.0000	.0000	.0000	.0010	.0091	.0350	.0820	.1376	.1892	.1214
	8	.0000	.0000	.0000	.0002	.0022	.0120	.0376	.0811	.1734	.1669
	9	.0000	.0000	.0000	.0000	.0004	.0033	.0139	.0386	.1284	.1855
	10	.0000	.0000	.0000	.0000	.0001	.0008	.0042	.0149	.0771	.1669
	11	.0000	.0000	.0000	.0000	.0000	.0001	.0010	.0046	.0374	.1214
	12	.0000	.0000	.0000	.0000	.0000	.0000	.0002	.0012	.0145	.0708
	13	.0000	.0000	.0000	.0000	.0000	.0000	.0000	.0002	.0045	.0327
	14	.0000	.0000	.0000	.0000	.0000	.0000	.0000	.0000	.0011	.0117
	15	.0000	.0000	.0000	.0000	.0000	.0000	.0000	.0000	.0002	.0031
	16	.0000	.0000	.0000	.0000	.0000	.0000	.0000	.0000	.0000	.0006
	17	.0000	.0000	.0000	.0000	.0000	.0000	.0000	.0000	.0000	.0001
	18	.0000	.0000	.0000	.0000	.0000	.0000	.0000	.0000	.0000	.0000
19	0	.8262	.6812	.3774	.1351	.0456	.0144	.0042	.0011	.0001	.0000
	1	.1586	.2642	.3774	.2852	.1529	.0685	.0268	.0093	.0008	.0000
	2	.0144	.0485	.1787	.2852	.2428	.1540	.0803	.0358	.0046	.0003
	3	.0008	.0056	.0533	.1796	.2428	.2182	.1517	.0869	.0175	.0018
	4	.0000	.0005	.0112	.0798	.1714	.2182	.2023	.1491	.0467	.0074
	5	.0000	.0000	.0018	.0266	.0907	.1636	.2023	.1916	.0933	.0222
	6	.0000	.0000	.0002	.0069	.0374	.0955	.1574	.1916	.1451	.0518
	7	.0000	.0000	.0000	.0014	.0122	.0443	.0974	.1525	.1797	.0961
	8	.0000	.0000	.0000	.0002	.0032	.0166	.0487	.0981	.1797	.1442
	9	.0000	.0000	.0000	.0000	.0007	.0051	.0198	.0514	.1464	.1762
	10	.0000	.0000	.0000	.0000	.0001	.0013	.0066	.0220	.0976	.1762
	11	.0000	.0000	.0000	.0000	.0000	.0003	.0018	.0077	.0532	.1442
	12	.0000	.0000	.0000	.0000	.0000	.0000	.0004	.0022	.0237	.0961
	13	.0000	.0000	.0000	.0000	.0000	.0000	.0001	.0005	.0085	.0518
	14	.0000	.0000	.0000	.0000	.0000	.0000	.0000	.0001	.0024	.0222
	15	.0000	.0000	.0000	.0000	.0000	.0000	.0000	.0000	.0005	.0074
	16	.0000	.0000	.0000	.0000	.0000	.0000	.0000	.0000	.0001	.0018
	17	.0000	.0000	.0000	.0000	.0000	.0000	.0000	.0000	.0000	.0003
	18	.0000	.0000	.0000	.0000	.0000	.0000	.0000	.0000	.0000	.0000
	19	.0000	.0000	.0000	.0000	.0000	.0000	.0000	.0000	.0000	.0000

Table A-3. **Table of binomial probability. (continued)**

n	x	.01	.02	.05	.10	p .15	.20	.25	.30	.40	.50
20	0	.8179	.6676	.3585	.1216	.0388	.0115	.0032	.0008	.0000	.0000
	1	.1652	.2725	.3774	.2702	.1368	.0576	.0211	.0068	.0005	.0000
	2	.0159	.0528	.1887	.2852	.2293	.1369	.0669	.0278	.0031	.0002
	3	.0010	.0065	.0596	.1901	.2428	.2054	.1339	.0716	.0123	.0011
	4	.0000	.0006	.0133	.0898	.1821	.2182	.1897	.1304	.0350	.0046
	5	.0000	.0000	.0022	.0319	.1028	.1746	.2023	.1789	.0746	.0148
	6	.0000	.0000	.0003	.0089	.0454	.1091	.1686	.1916	.1244	.0370
	7	.0000	.0000	.0000	.0020	.0160	.0545	.1124	.1643	.1659	.0739
	8	.0000	.0000	.0000	.0004	.0046	.0222	.0609	.1144	.1797	.1201
	9	.0000	.0000	.0000	.0001	.0011	.0074	.0271	.0654	.1597	.1602
	10	.0000	.0000	.0000	.0000	.0002	.0020	.0099	.0308	.1171	.1762
	11	.0000	.0000	.0000	.0000	.0000	.0005	.0030	.0120	.0710	.1602
	12	.0000	.0000	.0000	.0000	.0000	.0001	.0008	.0039	.0355	.1201
	13	.0000	.0000	.0000	.0000	.0000	.0000	.0002	.0010	.0146	.0739
	14	.0000	.0000	.0000	.0000	.0000	.0000	.0000	.0002	.0049	.0370
	15	.0000	.0000	.0000	.0000	.0000	.0000	.0000	.0000	.0013	.0148
	16	.0000	.0000	.0000	.0000	.0000	.0000	.0000	.0000	.0003	.0046
	17	.0000	.0000	.0000	.0000	.0000	.0000	.0000	.0000	.0000	.0011
	18	.0000	.0000	.0000	.0000	.0000	.0000	.0000	.0000	.0000	.0002
	19	.0000	.0000	.0000	.0000	.0000	.0000	.0000	.0000	.0000	.0000
	20	.0000	.0000	.0000	.0000	.0000	.0000	.0000	.0000	.0000	.0000

TABLE A-4
CUMULATIVE BINOMIAL PROBABILITY

In the example given with the explanation of Table A-3, you selected five clocks at random, and found the probability of three of them being set to the correct time. Cumulative probability looks at the probability of the given number *or less* (in this example, the probability of three or less; that is, three or two or one or zero).

The procedure is the same as with the noncumulative table. Look down the n column to 5; locate 3 in the adjacent group under x; and go across that row to the column headed .10. At that intersection, read the cumulative probability of 0.9995.

You could have obtained the same result from the noncumulative table (Table A-3) by finding the probability of three (.0081), of two (0.0729), of one (0.3280), and of zero (0.5905), and adding the probabilities. The cumulative table saves you the time of looking up separate probabilities and adding them.

Table A-4. Cumulative binomial probability

n	x	.01	.02	.05	.10	p .15	.20	.25	.30	.40	.50
1	0	0.9900	0.9800	0.9500	0.9000	0.8500	0.8000	0.7500	0.7000	0.6000	0.5000
	1	1.0000	1.0000	1.0000	1.0000	1.0000	1.0000	1.0000	1.0000	1.0000	1.0000
2	0	0.9801	0.9604	0.9025	0.8100	0.7225	0.6400	0.5625	0.4900	0.3600	0.2500
	1	0.9999	0.9996	0.9975	0.9900	0.9775	0.9600	0.9375	0.9100	0.8400	0.7500
	2	1.0000	1.0000	1.0000	1.0000	1.0000	1.0000	1.0000	1.0000	1.0000	1.0000
3	0	0.9703	0.9412	0.8574	0.7290	0.6141	0.5120	0.4219	0.3430	0.2160	0.1250
	1	0.9997	0.9988	0.9927	0.9720	0.9393	0.8960	0.8438	0.7840	0.6480	0.5000
	2	1.0000	1.0000	0.9999	0.9990	0.9966	0.9920	0.9844	0.9730	0.9360	0.8750
	3	1.0000	1.0000	1.0000	1.0000	1.0000	1.0000	1.0000	1.0000	1.0000	1.0000
4	0	0.9606	0.9224	0.8145	0.6561	0.5220	0.4096	0.3164	0.2401	0.1296	0.0625
	1	0.9994	0.9977	0.9860	0.9477	0.8905	0.8192	0.7383	0.6517	0.4752	0.3125
	2	1.0000	1.0000	0.9995	0.9963	0.9880	0.9728	0.9492	0.9163	0.8208	0.6875
	3	1.0000	1.0000	1.0000	0.9999	0.9995	0.9984	0.9961	0.9919	0.9744	0.9375
	4	1.0000	1.0000	1.0000	1.0000	1.0000	1.0000	1.0000	1.0000	1.0000	1.0000
5	0	0.9510	0.9039	0.7738	0.5905	0.4437	0.3277	0.2373	0.1681	0.0778	0.0313
	1	0.9990	0.9962	0.9774	0.9185	0.8352	0.7373	0.6328	0.5282	0.3370	0.1875
	2	1.0000	0.9999	0.9988	0.9914	0.9734	0.9421	0.8965	0.8369	0.6826	0.5000
	3	1.0000	1.0000	1.0000	0.9995	0.9978	0.9933	0.9844	0.9692	0.9130	0.8125
	4	1.0000	1.0000	1.0000	1.0000	0.9999	0.9997	0.9990	0.9976	0.9898	0.9688
	5	1.0000	1.0000	1.0000	1.0000	1.0000	1.0000	1.0000	1.0000	1.0000	1.0000
6	0	0.9415	0.8858	0.7351	0.5314	0.3771	0.2621	0.1780	0.1176	0.0467	0.0156
	1	0.9985	0.9943	0.9672	0.8857	0.7765	0.6554	0.5339	0.4202	0.2333	0.1094
	2	1.0000	0.9998	0.9978	0.9841	0.9527	0.9011	0.8306	0.7443	0.5443	0.3438
	3	1.0000	1.0000	0.9999	0.9987	0.9941	0.9830	0.9624	0.9295	0.8208	0.6563
	4	1.0000	1.0000	1.0000	0.9999	0.9996	0.9984	0.9954	0.9891	0.9590	0.8906
	5	1.0000	1.0000	1.0000	1.0000	1.0000	0.9999	0.9998	0.9993	0.9959	0.9844
	6	1.0000	1.0000	1.0000	1.0000	1.0000	1.0000	1.0000	1.0000	1.0000	1.0000
7	0	0.9321	0.8681	0.6983	0.4783	0.3206	0.2097	0.1335	0.0824	0.0280	0.0078
	1	0.9980	0.9921	0.9556	0.8503	0.7166	0.5767	0.4449	0.3294	0.1586	0.0625
	2	1.0000	0.9997	0.9962	0.9743	0.9262	0.8520	0.7564	0.6471	0.4199	0.2266
	3	1.0000	1.0000	0.9998	0.9973	0.9879	0.9667	0.9294	0.8740	0.7102	0.5000
	4	1.0000	1.0000	1.0000	0.9998	0.9988	0.9953	0.9871	0.9712	0.9037	0.7734
	5	1.0000	1.0000	1.0000	1.0000	0.9999	0.9996	0.9987	0.9962	0.9812	0.9375
	6	1.0000	1.0000	1.0000	1.0000	1.0000	1.0000	0.9999	0.9998	0.9984	0.9922
	7	1.0000	1.0000	1.0000	1.0000	1.0000	1.0000	1.0000	1.0000	1.0000	1.0000
8	0	0.9227	0.8508	0.6634	0.4305	0.2725	0.1678	0.1001	0.0576	0.0168	0.0039
	1	0.9973	0.9897	0.9428	0.8131	0.6572	0.5033	0.3671	0.2553	0.1064	0.0352
	2	0.9999	0.9996	0.9942	0.9619	0.8948	0.7969	0.6785	0.5518	0.3154	0.1445
	3	1.0000	1.0000	0.9996	0.9950	0.9786	0.9437	0.8862	0.8059	0.5941	0.3633
	4	1.0000	1.0000	1.0000	0.9996	0.9971	0.9896	0.9727	0.9420	0.8263	0.6367
	5	1.0000	1.0000	1.0000	1.0000	0.9998	0.9988	0.9958	0.9887	0.9502	0.8555
	6	1.0000	1.0000	1.0000	1.0000	1.0000	0.9999	0.9996	0.9987	0.9915	0.9648
	7	1.0000	1.0000	1.0000	1.0000	1.0000	1.0000	1.0000	0.9999	0.9993	0.9961
	8	1.0000	1.0000	1.0000	1.0000	1.0000	1.0000	1.0000	1.0000	1.0000	1.0000

Table A-4. Cumulative binomial probability. (continued)

n	x	.01	.02	.05	.10	p .15	.20	.25	.30	.40	.50
9	0	0.9135	0.8337	0.6302	0.3874	0.2316	0.1342	0.0751	0.0404	0.0101	0.0020
	1	0.9966	0.9869	0.9288	0.7748	0.5995	0.4362	0.3003	0.1960	0.0705	0.0195
	2	0.9999	0.9994	0.9916	0.9470	0.8591	0.7382	0.6007	0.4628	0.2318	0.0898
	3	1.0000	1.0000	0.9994	0.9917	0.9661	0.9144	0.8343	0.7297	0.4826	0.2539
	4	1.0000	1.0000	1.0000	0.9991	0.9944	0.9804	0.9511	0.9012	0.7334	0.5000
	5	1.0000	1.0000	1.0000	0.9999	0.9994	0.9969	0.9900	0.9747	0.9006	0.7461
	6	1.0000	1.0000	1.0000	1.0000	1.0000	0.9997	0.9987	0.9957	0.9750	0.9102
	7	1.0000	1.0000	1.0000	1.0000	1.0000	1.0000	0.9999	0.9996	0.9962	0.9805
	8	1.0000	1.0000	1.0000	1.0000	1.0000	1.0000	1.0000	1.0000	0.9997	0.9980
	9	1.0000	1.0000	1.0000	1.0000	1.0000	1.0000	1.0000	1.0000	1.0000	1.0000
10	0	0.9044	0.8171	0.5987	0.3487	0.1969	0.1074	0.0563	0.0282	0.0060	0.0010
	1	0.9957	0.9838	0.9139	0.7361	0.5443	0.3758	0.2440	0.1493	0.0464	0.0107
	2	0.9999	0.9991	0.9885	0.9298	0.8202	0.6778	0.5256	0.3828	0.1673	0.0547
	3	1.0000	1.0000	0.9990	0.9872	0.9500	0.8791	0.7759	0.6496	0.3823	0.1719
	4	1.0000	1.0000	0.9999	0.9984	0.9901	0.9672	0.9219	0.8497	0.6331	0.3770
	5	1.0000	1.0000	1.0000	0.9999	0.9986	0.9936	0.9803	0.9527	0.8338	0.6230
	6	1.0000	1.0000	1.0000	1.0000	0.9999	0.9991	0.9965	0.9894	0.9452	0.8281
	7	1.0000	1.0000	1.0000	1.0000	1.0000	0.9999	0.9996	0.9984	0.9877	0.9453
	8	1.0000	1.0000	1.0000	1.0000	1.0000	1.0000	1.0000	0.9999	0.9983	0.9893
	9	1.0000	1.0000	1.0000	1.0000	1.0000	1.0000	1.0000	1.0000	0.9999	0.9990
	10	1.0000	1.0000	1.0000	1.0000	1.0000	1.0000	1.0000	1.0000	1.0000	1.0000
11	0	0.8953	0.8007	0.5688	0.3138	0.1673	0.0859	0.0422	0.0198	0.0036	0.0005
	1	0.9948	0.9805	0.8981	0.6974	0.4922	0.3221	0.1971	0.1130	0.0302	0.0059
	2	0.9998	0.9988	0.9848	0.9104	0.7788	0.6174	0.4552	0.3127	0.1189	0.0327
	3	1.0000	1.0000	0.9984	0.9815	0.9306	0.8389	0.7133	0.5696	0.2963	0.1133
	4	1.0000	1.0000	0.9999	0.9972	0.9841	0.9496	0.8854	0.7897	0.5328	0.2744
	5	1.0000	1.0000	1.0000	0.9997	0.9973	0.9883	0.9657	0.9218	0.7535	0.5000
	6	1.0000	1.0000	1.0000	1.0000	0.9997	0.9980	0.9924	0.9784	0.9006	0.7256
	7	1.0000	1.0000	1.0000	1.0000	1.0000	0.9998	0.9988	0.9957	0.9707	0.8867
	8	1.0000	1.0000	1.0000	1.0000	1.0000	1.0000	0.9999	0.9994	0.9941	0.9673
	9	1.0000	1.0000	1.0000	1.0000	1.0000	1.0000	1.0000	1.0000	0.9993	0.9941
	10	1.0000	1.0000	1.0000	1.0000	1.0000	1.0000	1.0000	1.0000	1.0000	0.9995
	11	1.0000	1.0000	1.0000	1.0000	1.0000	1.0000	1.0000	1.0000	1.0000	1.0000
12	0	0.8864	0.7847	0.5404	0.2824	0.1422	0.0687	0.0317	0.0138	0.0022	0.0002
	1	0.9938	0.9769	0.8816	0.6590	0.4435	0.2749	0.1584	0.0850	0.0196	0.0032
	2	0.9998	0.9985	0.9804	0.8891	0.7358	0.5583	0.3907	0.2528	0.0834	0.0193
	3	1.0000	0.9999	0.9978	0.9744	0.9078	0.7946	0.6488	0.4925	0.2253	0.0730
	4	1.0000	1.0000	0.9998	0.9957	0.9761	0.9274	0.8424	0.7237	0.4382	0.1938
	5	1.0000	1.0000	1.0000	0.9995	0.9954	0.9806	0.9456	0.8822	0.6652	0.3872
	6	1.0000	1.0000	1.0000	0.9999	0.9993	0.9961	0.9857	0.9614	0.8418	0.6128
	7	1.0000	1.0000	1.0000	1.0000	0.9999	0.9994	0.9972	0.9905	0.9427	0.8062
	8	1.0000	1.0000	1.0000	1.0000	1.0000	0.9999	0.9996	0.9983	0.9847	0.9270
	9	1.0000	1.0000	1.0000	1.0000	1.0000	1.0000	1.0000	0.9998	0.9972	0.9807
	10	1.0000	1.0000	1.0000	1.0000	1.0000	1.0000	1.0000	1.0000	0.9997	0.9968
	11	1.0000	1.0000	1.0000	1.0000	1.0000	1.0000	1.0000	1.0000	1.0000	0.9998
	12	1.0000	1.0000	1.0000	1.0000	1.0000	1.0000	1.0000	1.0000	1.0000	1.0000

Table A-4.   Cumulative binomial probability. (continued)

n	x	.01	.02	.05	.10	p .15	.20	.25	.30	.40	.50
13	0	0.8775	0.7690	0.5133	0.2542	0.1209	0.0550	0.0238	0.0097	0.0013	0.0001
	1	0.9928	0.9730	0.8646	0.6213	0.3983	0.2336	0.1267	0.0637	0.0126	0.0017
	2	0.9997	0.9980	0.9755	0.8661	0.6920	0.5017	0.3326	0.2025	0.0579	0.0112
	3	1.0000	0.9999	0.9969	0.9658	0.8820	0.7473	0.5843	0.4206	0.1686	0.0461
	4	1.0000	1.0000	0.9997	0.9935	0.9658	0.9009	0.7940	0.6543	0.3530	0.1334
	5	1.0000	1.0000	1.0000	0.9991	0.9925	0.9700	0.9198	0.8346	0.5744	0.2905
	6	1.0000	1.0000	1.0000	0.9999	0.9987	0.9930	0.9757	0.9376	0.7712	0.5000
	7	1.0000	1.0000	1.0000	1.0000	0.9998	0.9988	0.9944	0.9818	0.9023	0.7095
	8	1.0000	1.0000	1.0000	1.0000	1.0000	0.9998	0.9990	0.9960	0.9679	0.8666
	9	1.0000	1.0000	1.0000	1.0000	1.0000	1.0000	0.9999	0.9993	0.9922	0.9539
	10	1.0000	1.0000	1.0000	1.0000	1.0000	1.0000	1.0000	0.9999	0.9987	0.9888
	11	1.0000	1.0000	1.0000	1.0000	1.0000	1.0000	1.0000	1.0000	0.9999	0.9983
	12	1.0000	1.0000	1.0000	1.0000	1.0000	1.0000	1.0000	1.0000	1.0000	0.9999
	13	1.0000	1.0000	1.0000	1.0000	1.0000	1.0000	1.0000	1.0000	1.0000	1.0000
14	0	0.8687	0.7536	0.4877	0.2288	0.1028	0.0440	0.0178	0.0068	0.0008	0.0001
	1	0.9916	0.9690	0.8470	0.5846	0.3567	0.1979	0.1010	0.0475	0.0081	0.0009
	2	0.9997	0.9975	0.9699	0.8416	0.6479	0.4481	0.2811	0.1608	0.0398	0.0065
	3	1.0000	0.9999	0.9958	0.9559	0.8535	0.6982	0.5213	0.3552	0.1243	0.0287
	4	1.0000	1.0000	0.9996	0.9908	0.9533	0.8702	0.7415	0.5842	0.2793	0.0898
	5	1.0000	1.0000	1.0000	0.9985	0.9885	0.9561	0.8883	0.7805	0.4859	0.2120
	6	1.0000	1.0000	1.0000	0.9998	0.9978	0.9884	0.9617	0.9067	0.6925	0.3953
	7	1.0000	1.0000	1.0000	1.0000	0.9997	0.9976	0.9897	0.9685	0.8499	0.6047
	8	1.0000	1.0000	1.0000	1.0000	1.0000	0.9996	0.9978	0.9917	0.9417	0.7880
	9	1.0000	1.0000	1.0000	1.0000	1.0000	1.0000	0.9997	0.9983	0.9825	0.9102
	10	1.0000	1.0000	1.0000	1.0000	1.0000	1.0000	1.0000	0.9998	0.9961	0.9713
	11	1.0000	1.0000	1.0000	1.0000	1.0000	1.0000	1.0000	1.0000	0.9994	0.9935
	12	1.0000	1.0000	1.0000	1.0000	1.0000	1.0000	1.0000	1.0000	0.9999	0.9991
	13	1.0000	1.0000	1.0000	1.0000	1.0000	1.0000	1.0000	1.0000	1.0000	0.9999
	14	1.0000	1.0000	1.0000	1.0000	1.0000	1.0000	1.0000	1.0000	1.0000	1.0000
15	0	0.8601	0.7386	0.4633	0.2059	0.0874	0.0352	0.0134	0.0047	0.0005	0.0000
	1	0.9904	0.9647	0.8290	0.5490	0.3186	0.1671	0.0802	0.0353	0.0052	0.0005
	2	0.9996	0.9970	0.9638	0.8159	0.6042	0.3980	0.2361	0.1268	0.0271	0.0037
	3	1.0000	0.9998	0.9945	0.9444	0.8227	0.6482	0.4613	0.2969	0.0905	0.0176
	4	1.0000	1.0000	0.9994	0.9873	0.9383	0.8358	0.6865	0.5155	0.2173	0.0592
	5	1.0000	1.0000	0.9999	0.9977	0.9832	0.9389	0.8516	0.7216	0.4032	0.1509
	6	1.0000	1.0000	1.0000	0.9997	0.9964	0.9819	0.9434	0.8689	0.6098	0.3036
	7	1.0000	1.0000	1.0000	1.0000	0.9994	0.9958	0.9827	0.9500	0.7869	0.5000
	8	1.0000	1.0000	1.0000	1.0000	0.9999	0.9992	0.9958	0.9848	0.9050	0.6964
	9	1.0000	1.0000	1.0000	1.0000	1.0000	0.9999	0.9992	0.9963	0.9662	0.8491
	10	1.0000	1.0000	1.0000	1.0000	1.0000	1.0000	0.9999	0.9993	0.9907	0.9408
	11	1.0000	1.0000	1.0000	1.0000	1.0000	1.0000	1.0000	0.9999	0.9981	0.9824
	12	1.0000	1.0000	1.0000	1.0000	1.0000	1.0000	1.0000	1.0000	0.9997	0.9963
	13	1.0000	1.0000	1.0000	1.0000	1.0000	1.0000	1.0000	1.0000	1.0000	0.9995
	14	1.0000	1.0000	1.0000	1.0000	1.0000	1.0000	1.0000	1.0000	1.0000	1.0000
	15	1.0000	1.0000	1.0000	1.0000	1.0000	1.0000	1.0000	1.0000	1.0000	1.0000

Table A-4. Cumulative binomial probability. (continued)

n	x	.01	.02	.05	.10	.15	.20	.25	.30	.40	.50
16	0	0.8515	0.7238	0.4401	0.1853	0.0743	0.0281	0.0100	0.0033	0.0003	0.0000
	1	0.9891	0.9601	0.8108	0.5147	0.2839	0.1407	0.0635	0.0261	0.0033	0.0003
	2	0.9995	0.9963	0.9571	0.7892	0.5614	0.3518	0.1971	0.0994	0.0183	0.0021
	3	1.0000	0.9998	0.9930	0.9316	0.7899	0.5981	0.4050	0.2459	0.0651	0.0106
	4	1.0000	1.0000	0.9991	0.9830	0.9209	0.7982	0.6302	0.4499	0.1666	0.0384
	5	1.0000	1.0000	0.9999	0.9967	0.9765	0.9183	0.8103	0.6598	0.3288	0.1051
	6	1.0000	1.0000	1.0000	0.9995	0.9944	0.9733	0.9204	0.8247	0.5272	0.2272
	7	1.0000	1.0000	1.0000	0.9999	0.9989	0.9930	0.9729	0.9256	0.7161	0.4018
	8	1.0000	1.0000	1.0000	1.0000	0.9998	0.9985	0.9925	0.9743	0.8577	0.5982
	9	1.0000	1.0000	1.0000	1.0000	1.0000	0.9998	0.9984	0.9929	0.9417	0.7728
	10	1.0000	1.0000	1.0000	1.0000	1.0000	1.0000	0.9997	0.9984	0.9809	0.8949
	11	1.0000	1.0000	1.0000	1.0000	1.0000	1.0000	1.0000	0.9997	0.9951	0.9616
	12	1.0000	1.0000	1.0000	1.0000	1.0000	1.0000	1.0000	1.0000	0.9991	0.9894
	13	1.0000	1.0000	1.0000	1.0000	1.0000	1.0000	1.0000	1.0000	0.9999	0.9979
	14	1.0000	1.0000	1.0000	1.0000	1.0000	1.0000	1.0000	1.0000	1.0000	0.9997
	15	1.0000	1.0000	1.0000	1.0000	1.0000	1.0000	1.0000	1.0000	1.0000	1.0000
	16	1.0000	1.0000	1.0000	1.0000	1.0000	1.0000	1.0000	1.0000	1.0000	1.0000
17	0	0.8429	0.7093	0.4181	0.1668	0.0631	0.0225	0.0075	0.0023	0.0002	0.0000
	1	0.9877	0.9554	0.7922	0.4818	0.2525	0.1182	0.0501	0.0193	0.0021	0.0001
	2	0.9994	0.9956	0.9497	0.7618	0.5198	0.3096	0.1637	0.0774	0.0123	0.0012
	3	1.0000	0.9997	0.9912	0.9174	0.7556	0.5489	0.3530	0.2019	0.0464	0.0064
	4	1.0000	1.0000	0.9988	0.9779	0.9013	0.7582	0.5739	0.3887	0.1260	0.0245
	5	1.0000	1.0000	0.9999	0.9953	0.9681	0.8943	0.7653	0.5968	0.2639	0.0717
	6	1.0000	1.0000	1.0000	0.9992	0.9917	0.9623	0.8929	0.7752	0.4478	0.1662
	7	1.0000	1.0000	1.0000	0.9999	0.9983	0.9891	0.9598	0.8954	0.6405	0.3145
	8	1.0000	1.0000	1.0000	1.0000	0.9997	0.9974	0.9876	0.9597	0.8011	0.5000
	9	1.0000	1.0000	1.0000	1.0000	1.0000	0.9995	0.9969	0.9873	0.9081	0.6855
	10	1.0000	1.0000	1.0000	1.0000	1.0000	0.9999	0.9994	0.9968	0.9652	0.8338
	11	1.0000	1.0000	1.0000	1.0000	1.0000	1.0000	0.9999	0.9993	0.9894	0.9283
	12	1.0000	1.0000	1.0000	1.0000	1.0000	1.0000	1.0000	0.9999	0.9975	0.9755
	13	1.0000	1.0000	1.0000	1.0000	1.0000	1.0000	1.0000	1.0000	0.9995	0.9936
	14	1.0000	1.0000	1.0000	1.0000	1.0000	1.0000	1.0000	1.0000	0.9999	0.9988
	15	1.0000	1.0000	1.0000	1.0000	1.0000	1.0000	1.0000	1.0000	1.0000	0.9999
	16	1.0000	1.0000	1.0000	1.0000	1.0000	1.0000	1.0000	1.0000	1.0000	1.0000
	17	1.0000	1.0000	1.0000	1.0000	1.0000	1.0000	1.0000	1.0000	1.0000	1.0000

Table A-4. Cumulative binomial probability. (continued)

n	x	.01	.02	.05	.10	.15	.20	.25	.30	.40	.50
18	0	0.8345	0.6951	0.3972	0.1501	0.0536	0.0180	0.0056	0.0016	0.0001	0.0000
	1	0.9862	0.9505	0.7735	0.4503	0.2241	0.0991	0.0395	0.0142	0.0013	0.0001
	2	0.9993	0.9948	0.9419	0.7338	0.4797	0.2713	0.1353	0.0600	0.0082	0.0007
	3	1.0000	0.9996	0.9891	0.9018	0.7202	0.5010	0.3057	0.1646	0.0328	0.0038
	4	1.0000	1.0000	0.9985	0.9718	0.8794	0.7164	0.5187	0.3327	0.0942	0.0154
	5	1.0000	1.0000	0.9998	0.9936	0.9581	0.8671	0.7175	0.5344	0.2088	0.0481
	6	1.0000	1.0000	1.0000	0.9988	0.9882	0.9487	0.8610	0.7217	0.3743	0.1189
	7	1.0000	1.0000	1.0000	0.9998	0.9973	0.9837	0.9431	0.8593	0.5634	0.2403
	8	1.0000	1.0000	1.0000	1.0000	0.9995	0.9957	0.9807	0.9404	0.7368	0.4073
	9	1.0000	1.0000	1.0000	1.0000	0.9999	0.9991	0.9946	0.9790	0.8653	0.5927
	10	1.0000	1.0000	1.0000	1.0000	1.0000	0.9998	0.9988	0.9939	0.9424	0.7597
	11	1.0000	1.0000	1.0000	1.0000	1.0000	1.0000	0.9998	0.9986	0.9797	0.8811
	12	1.0000	1.0000	1.0000	1.0000	1.0000	1.0000	1.0000	0.9997	0.9942	0.9519
	13	1.0000	1.0000	1.0000	1.0000	1.0000	1.0000	1.0000	1.0000	0.9987	0.9846
	14	1.0000	1.0000	1.0000	1.0000	1.0000	1.0000	1.0000	1.0000	0.9998	0.9962
	15	1.0000	1.0000	1.0000	1.0000	1.0000	1.0000	1.0000	1.0000	1.0000	0.9993
	16	1.0000	1.0000	1.0000	1.0000	1.0000	1.0000	1.0000	1.0000	1.0000	0.9999
	17	1.0000	1.0000	1.0000	1.0000	1.0000	1.0000	1.0000	1.0000	1.0000	1.0000
	18	1.0000	1.0000	1.0000	1.0000	1.0000	1.0000	1.0000	1.0000	1.0000	1.0000
19	0	0.8262	0.6812	0.3774	0.1351	0.0456	0.0144	0.0042	0.0011	0.0001	0.0000
	1	0.9847	0.9454	0.7547	0.4203	0.1985	0.0829	0.0310	0.0104	0.0008	0.0000
	2	0.9991	0.9939	0.9335	0.7054	0.4413	0.2369	0.1113	0.0462	0.0055	0.0004
	3	1.0000	0.9995	0.9868	0.8850	0.6841	0.4551	0.2631	0.1332	0.0230	0.0022
	4	1.0000	1.0000	0.9980	0.9648	0.8556	0.6733	0.4654	0.2822	0.0696	0.0096
	5	1.0000	1.0000	0.9998	0.9914	0.9463	0.8369	0.6678	0.4739	0.1629	0.0318
	6	1.0000	1.0000	1.0000	0.9983	0.9837	0.9324	0.8251	0.6655	0.3081	0.0835
	7	1.0000	1.0000	1.0000	0.9997	0.9959	0.9767	0.9225	0.8180	0.4878	0.1796
	8	1.0000	1.0000	1.0000	1.0000	0.9992	0.9933	0.9713	0.9161	0.6675	0.3238
	9	1.0000	1.0000	1.0000	1.0000	0.9999	0.9984	0.9911	0.9674	0.8139	0.5000
	10	1.0000	1.0000	1.0000	1.0000	1.0000	0.9997	0.9977	0.9895	0.9115	0.6762
	11	1.0000	1.0000	1.0000	1.0000	1.0000	0.9999	0.9995	0.9972	0.9648	0.8204
	12	1.0000	1.0000	1.0000	1.0000	1.0000	1.0000	0.9999	0.9994	0.9884	0.9165
	13	1.0000	1.0000	1.0000	1.0000	1.0000	1.0000	1.0000	0.9999	0.9969	0.9682
	14	1.0000	1.0000	1.0000	1.0000	1.0000	1.0000	1.0000	1.0000	0.9994	0.9904
	15	1.0000	1.0000	1.0000	1.0000	1.0000	1.0000	1.0000	1.0000	0.9999	0.9978
	16	1.0000	1.0000	1.0000	1.0000	1.0000	1.0000	1.0000	1.0000	1.0000	0.9996
	17	1.0000	1.0000	1.0000	1.0000	1.0000	1.0000	1.0000	1.0000	1.0000	1.0000
	18	1.0000	1.0000	1.0000	1.0000	1.0000	1.0000	1.0000	1.0000	1.0000	1.0000
	19	1.0000	1.0000	1.0000	1.0000	1.0000	1.0000	1.0000	1.0000	1.0000	1.0000

Table A-4. Cumulative binomial probability. (continued)

n	x	.01	.02	.05	.10	p .15	.20	.25	.30	.40	.50
20	0	0.8179	0.6676	0.3585	0.1216	0.0388	0.0115	0.0032	0.0008	0.0000	0.0000
	1	0.9831	0.9401	0.7358	0.3917	0.1756	0.0692	0.0243	0.0076	0.0005	0.0000
	2	0.9990	0.9929	0.9245	0.6769	0.4049	0.2061	0.0913	0.0355	0.0036	0.0002
	3	1.0000	0.9994	0.9841	0.8670	0.6477	0.4114	0.2252	0.1071	0.0160	0.0013
	4	1.0000	1.0000	0.9974	0.9568	0.8298	0.6296	0.4148	0.2375	0.0510	0.0059
	5	1.0000	1.0000	0.9997	0.9887	0.9327	0.8042	0.6172	0.4164	0.1256	0.0207
	6	1.0000	1.0000	1.0000	0.9976	0.9781	0.9133	0.7858	0.6080	0.2500	0.0577
	7	1.0000	1.0000	1.0000	0.9996	0.9941	0.9679	0.8982	0.7723	0.4159	0.1316
	8	1.0000	1.0000	1.0000	0.9999	0.9987	0.9900	0.9591	0.8867	0.5956	0.2517
	9	1.0000	1.0000	1.0000	1.0000	0.9998	0.9974	0.9861	0.9520	0.7553	0.4119
	10	1.0000	1.0000	1.0000	1.0000	1.0000	0.9994	0.9961	0.9829	0.8725	0.5881
	11	1.0000	1.0000	1.0000	1.0000	1.0000	0.9999	0.9991	0.9949	0.9435	0.7483
	12	1.0000	1.0000	1.0000	1.0000	1.0000	1.0000	0.9998	0.9987	0.9790	0.8684
	13	1.0000	1.0000	1.0000	1.0000	1.0000	1.0000	1.0000	0.9997	0.9935	0.9423
	14	1.0000	1.0000	1.0000	1.0000	1.0000	1.0000	1.0000	1.0000	0.9984	0.9793
	15	1.0000	1.0000	1.0000	1.0000	1.0000	1.0000	1.0000	1.0000	0.9997	0.9941
	16	1.0000	1.0000	1.0000	1.0000	1.0000	1.0000	1.0000	1.0000	1.0000	0.9987
	17	1.0000	1.0000	1.0000	1.0000	1.0000	1.0000	1.0000	1.0000	1.0000	0.9998
	18	1.0000	1.0000	1.0000	1.0000	1.0000	1.0000	1.0000	1.0000	1.0000	1.0000
	19	1.0000	1.0000	1.0000	1.0000	1.0000	1.0000	1.0000	1.0000	1.0000	1.0000
	20	1.0000	1.0000	1.0000	1.0000	1.0000	1.0000	1.0000	1.0000	1.0000	1.0000

INDEX

INDEX